Statistics

A course for
A level Mathematics

Book 1

D0532322

M. E. M. Jones

Head of the Mathematics Department at the Howardian High School, Cardiff

Schofield & Sims Ltd   Huddersfield

0 7217 2360 8
0 7217 2364 0 Net Edition

First printed 1988

Acknowledgements

The author and publishers are grateful for permission to use
questions from past G.C.E. examinations. These are
acknowledged as follows:

> Joint Matriculation Board (JMB)
> Welsh Joint Education Committee (WJEC)

Grateful acknowledgement is also due to the Joint
Matriculation Board for permission to reproduce the
statistical tables on pages 215 to 217 from their booklet
S159(A).

The Examining Boards whose questions are reproduced bear
no responsibility whatever for the answers to examination
questions given here, which are the sole responsibility of the
author.

Designed by Graphic Art Concepts, Leeds
Printed in England by The Bath Press, Avon

Author's note

The principal purpose of this book is to provide a straightforward text to cover the statistics content of the Advanced Level Mathematics (Pure and Applied Mathematics) syllabus of the Joint Matriculation Board. Together with Book 2 it also covers the statistics content of the syllabuses in Advanced Level Mathematics (Pure Mathematics with Statistics) and Further Mathematics (Pure and Applied Mathematics) of the Joint Matriculation Board, and in Advanced Level Mathematics and Applied Mathematics of the Welsh Joint Education Committee. The book aims to give a sound introduction to probabilistic and statistical concepts which are relevant to other school courses at both AS and A level, and also to introductory courses at colleges and universities.

Each chapter contains numerous worked examples to illustrate the methods by which the relevant theory is applied, followed by routine exercises to consolidate progress. At the end of each chapter is a miscellaneous exercise containing some more demanding questions most of which are taken from past papers of the Joint Matriculation Board (JMB) and Welsh Joint Education Committee (WJEC).

I wish to express my grateful thanks to a number of people who have assisted in the production of this book. In particular, to Dr. I. G. Evans, for reading the whole book in the early stages of its production and for making many helpful comments. Thanks also to the staff of Schofield and Sims for their invaluable help. Finally, thanks to my wife, Joan, for her help and encouragement and to my son, Christopher, for help with computing and word-processing.

M. E. M. Jones

Contents

iv

Notation

S	the sample space, the set of all possible outcomes of a random experiment
$A, B, C, \ldots$	subsets of S, events $A, B, C, \ldots$
$P(A)$	the probability of the event A
$P(A \mid B)$	the conditional probability of the event A given that the event B has occurred
$X, Y, Z, \ldots$	random variables $X, Y, Z, \ldots$
x_i	an arbitrary value of a discrete random variable X
p_i	the probability $P(X = x_i)$
$p(x_i)$	the value of the probability function p of a discrete random variable X
x	an arbitrary value of a continuous random variable X
$f(x)$	the value of the probability density function f of a continuous random variable X
$F(x)$	the value of the cumulative distribution function F of a continuous random variable X
$E[X]$	the expected value or expectation of a random variable X
$E[g(X)]$	the expected value of $g(X)$
$V[X]$	the variance of a random variable X
$SD[X]$	the standard deviation of a random variable X
μ	a population mean
σ^2	a population variance
σ	a population standard deviation
$\bar{x}$	a sample mean
s^2	a sample unbiased estimate of a population variance
$\sim$	is distributed as
$B(n, p)$	the binomial distribution with index n and probability parameter p
$Po(\alpha)$	the Poisson distribution with mean α
$U(a, b)$	the continuous uniform distribution over the interval (a, b)
$N(\mu, \sigma^2)$	the normal distribution with mean μ and variance σ^2
Z	the random variable having the distribution $N(0, 1)$
ϕ	the probability density function of Z
Φ	the cumulative distribution function of Z

Introduction

Statistics is a branch of applied mathematics which is concerned with the collection, organisation, representation, analysis and interpretation of data. The word 'statistics' is also used to refer to the set of data itself.

Historical note

The origins of the modern science of statistics may be traced to two disparate areas of investigation—political states and games of chance. The word 'statistics' was first used in the eighteenth century to refer to facts relating to a political state, i.e. information about such things as population, food production and industrial output. As an increasing proportion of this information was expressed in numerical form, the new word acquired the quantitative connotation it now has. Somewhat earlier, mathematicians, including Pascal (1623–1662), Fermat (1601–1665), Bernoulli (1654–1705) and de Moivre (1667–1754), had studied problems associated with games of chance, and as a result of their researches the foundations of a theory of probability had been constructed. Later, practical problems in subjects as diverse as astronomy and agriculture provided the stimuli for developments in statistical theory in which probability played an increasingly important part. Laplace (1749–1827) and Gauss (1777–1855) were two notable mathematicians who made significant contributions which led to a wider application of statistical methods.

Nowadays, the power of general methods based on probability concepts is recognised and these methods are applied in many widely different fields of study.

Descriptive and inferential statistics

In a statistical investigation, raw data (i.e. unorganised items of information) are collected and then classified; the classified data may be displayed in a wide variety of tables and diagrams for ease of comprehension. The reader will probably be familiar with bar charts and pie diagrams such as those below, which illustrate some aspects of the progress of an industrial company.

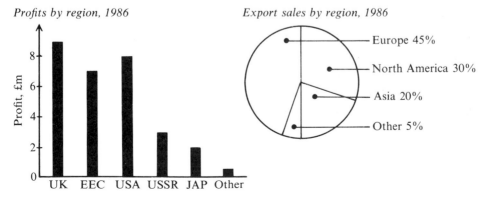

Profits by region, 1986

Export sales by region, 1986

Europe 45%

North America 30%

Asia 20%

Other 5%

1

In the bar chart, the length of each bar is proportional to the profit made by the company in £millions in the region indicated. In the pie diagram, the area of each sector is proportional to the percentage of the total exports of the company to the region shown.

Sometimes the calculation of averages helps to focus attention on specific facets of the performance of the company over a period of time; the table below shows the profit per employee for five successive years.

Year	1982	1983	1984	1985	1986
Profit per employee	£2099	£2512	£2993	£3258	£3492

An investigation in which data are organised into a form from which patterns may be discerned, diagrams drawn, averages calculated and reports drafted comes under the heading of *descriptive statistics*. Such investigations generally involve fairly simple techniques.

The second type of investigation, which involves statistical inference, demands methods of greater mathematical complexity. Sampling is fundamental to *inferential statistics* as it is usually impractical to obtain every possible measurement or observation of the process under investigation (the totality of measurements or observations is called a population). Therefore a representative sample (a subset of the population) is studied instead and properties of the population are inferred from the properties of the sample. For example, if 1000 people on the electoral roll are asked about their voting intentions in a forthcoming election, then their answers may be regarded as a sample from the population consisting of the voting intentions of all of the electorate. Using the sample data and a mathematical model of the situation, a prediction may be made about the whole population. On the basis of the opinion poll, a newspaper headline might proclaim:

Number 3882   Friday 8th May 1987   35p

POLL SAYS 38% WILL VOTE LABOUR

Using the same results a statistician would make a more cautious statement such as:

> "I can state with 95% confidence that the proportion of the electorate that intend to vote Labour in the next election lies between 35% and 41%."

There are two important points to note. Whenever an inference is made from the results of a statistical experiment, a numerical statement of the degree of certainty with which the inference is made is an essential feature of that inference; and in order that the conclusions should be reliable, it is vital that the sample is representative of the whole population. The former requires a careful application of probabilistic theory, whilst the selection of representative samples is one of the practical problems associated with the efficient design of statistical experiments.

This constitutes a very brief introduction to the science of statistics which is applied in a wide variety of fields such as agriculture, business studies, chemistry, economics, genetics, industrial quality control, military strategy, physics, psychology and sociology.

Chapter 1

Descriptive statistics

This chapter is mainly concerned with the organisation and representation of data in tabular and diagrammatic form. In statistics, data is collected in the form of measurements and observations of certain variables, e.g. the height of a person, the time of a runner in a race, the shoe size of a person, the number of goals scored in a league football match, the religious faith of a person. The first four of these variables are numerically valued and such variables may be divided into two broad classes, viz. *continuous* and *discrete*. If a variable may take any value in an interval, then it is said to be continuous in that interval; a variable which is not continuous in any interval is said to be discrete. Thus, the height of a person and the time of a runner in a race are examples of continuous variables, whilst the shoe size of a person and the number of goals scored in a football match are examples of discrete variables.

1.1 Frequency distributions

Example 1

Find and illustrate the frequency distribution of the shoe sizes of 100 men recorded below.

8	9	$6\frac{1}{2}$	11	8	$8\frac{1}{2}$	$8\frac{1}{2}$	10	$8\frac{1}{2}$	8	11	8	$8\frac{1}{2}$
9	$10\frac{1}{2}$	9	10	8	8	$6\frac{1}{2}$	$9\frac{1}{2}$	7	9	$9\frac{1}{2}$	6	$8\frac{1}{2}$
$8\frac{1}{2}$	9	$9\frac{1}{2}$	11	$10\frac{1}{2}$	10	$6\frac{1}{2}$	$7\frac{1}{2}$	10	7	7	8	$8\frac{1}{2}$
9	$7\frac{1}{2}$	$7\frac{1}{2}$	$8\frac{1}{2}$	$8\frac{1}{2}$	8	$7\frac{1}{2}$	9	$8\frac{1}{2}$	9	9	9	$9\frac{1}{2}$
$10\frac{1}{2}$	10	$9\frac{1}{2}$	$9\frac{1}{2}$	10	$7\frac{1}{2}$	$7\frac{1}{2}$	$10\frac{1}{2}$	8	9	9	9	9
$9\frac{1}{2}$	$8\frac{1}{2}$	$8\frac{1}{2}$	9	7	$8\frac{1}{2}$	10	7	$8\frac{1}{2}$	$8\frac{1}{2}$	9	9	$9\frac{1}{2}$
10	10	$9\frac{1}{2}$	$9\frac{1}{2}$	$10\frac{1}{2}$	$9\frac{1}{2}$	6	$7\frac{1}{2}$	8	$8\frac{1}{2}$	9	$10\frac{1}{2}$	$8\frac{1}{2}$
11	10	10	$9\frac{1}{2}$	$9\frac{1}{2}$	11	$10\frac{1}{2}$	11	$10\frac{1}{2}$				

It is difficult to discern any pattern in this information when it is presented in the above form, so an alternative method of presentation is often used to display the data in a more meaningful way. The *frequency* of each shoe size (i.e. the number of times each shoe size occurs in the list above) is calculated and the results recorded in **Table 1**, which is called the *frequency distribution* of the shoe sizes.

Table 1

Shoe size	6	$6\frac{1}{2}$	7	$7\frac{1}{2}$	8	$8\frac{1}{2}$	9	$9\frac{1}{2}$	10	$10\frac{1}{2}$	11
Frequency	2	3	5	7	10	17	18	13	11	8	6

The most appropriate method of illustrating the frequency distribution of a discrete variable is by means of a (*vertical*) *line diagram*, in which the frequency of a particular value of the variable is represented by the length of a 'vertical' line.

Line diagrams

The following line diagram represents the frequency distribution of the shoe sizes given in **Table 1**.

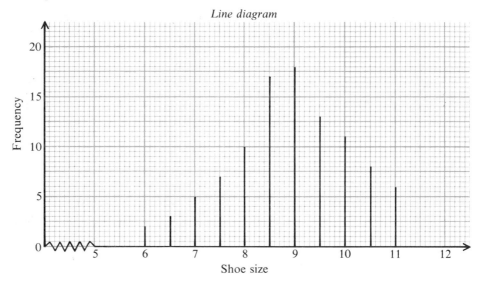

Line diagram

Grouped frequency distributions

Example 2

Find and illustrate a grouped frequency distribution for the heights (measured to the nearest cm) of 50 men recorded below.

172	181	166	173	174	178	175	170	180	182	182	171	172
172	173	177	176	175	176	160	185	199	171	177	179	178
178	189	190	187	175	177	172	172	168	181	183	186	190
191	172	175	181	176	181	182	172	176	178	179

In this example, the frequency distribution does not display the overall pattern very clearly, as the following extract shows.

Height (cm)	170	171	172	173	174	175	176	177	178	179
Frequency	1	2	7	2	1	4	4	3	4	2

In such cases it is usual to group the data into classes, sacrificing some detail to obtain a clearer overall picture. Suppose the data are grouped into eight classes, viz. 160–164, 165–169, 170–174, 175–179, 180–184, 185–189, 190–194, 195–199. The resulting *grouped frequency distribution* is shown in **Table 2**.

Table 2

Height (cm)	160–164	165–169	170–174	175–179	180–184	185–189	190–194	195–199
Frequency	1	2	13	17	9	4	3	1

The most appropriate method of illustrating a grouped frequency distribution is by means of a histogram. Before describing how a histogram is constructed, it is necessary to define some standard terms relating to grouped frequency distributions of continuous variables.

The *class limits* are the smallest and largest nominal values of the variable in the class; the class limits of the first class in the above grouped frequency distribution are 160 cm and 164 cm.

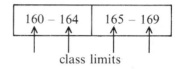

The *lower class boundary* of a class is the lowest value of the variable which, after it has been rounded, could be placed in the class; since the data have been rounded to the nearest whole number, the lower class boundary of the first class is 159.5 cm.

The *upper class boundary* of a class is the lower class boundary of the next higher class, so the upper class boundary of the first class is 164.5 cm.

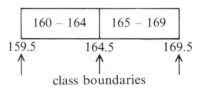

The *class width* is the difference between the upper and lower class boundaries; in this example the width of each class is 5 cm.

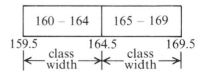

The *class mark* or *mid-interval value* is the value midway between the class boundaries; the mid-interval of the first class is 162 cm.

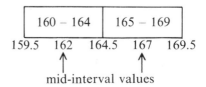

5

Histograms

In a *histogram*, each class frequency is represented by the area of a rectangle whose base extends from the lower class boundary to the upper class boundary; thus the width of the base is equal to the class width and the height of the rectangle may be found by dividing the frequency by the class width.

In the table below, each entry in the last column gives the height of the rectangle for the corresponding class in the grouped frequency distribution of the heights shown in **Table 2**.

Class	Frequency	Class boundaries		Class width	Frequency/class width
160–164	1	159.5	164.5	5	0.2
165–169	2	164.5	169.5	5	0.4
170–174	13	169.5	174.5	5	2.6
175–179	17	174.5	179.5	5	3.4
180–184	9	179.5	184.5	5	1.8
185–189	4	184.5	189.5	5	0.8
190–194	3	189.5	194.5	5	0.6
195–199	1	194.5	199.5	5	0.2

The 'vertical' axis should be labelled 'frequency/class width' and it must start at zero. In order to emphasise the fact that area represents frequency in a histogram, it is advisable to indicate the scale by means of a small square or rectangle, the area of which represents a certain frequency.

The histogram below illustrates the grouped frequency distribution given in **Table 2**.

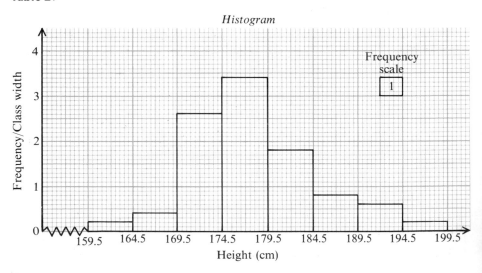

Histogram

Alternatively, when the class intervals of a grouped frequency distribution are all of equal width, the histogram may be constructed in the following manner. Draw a bar chart with the base of the bar for each class extending from the lower class boundary to the upper class boundary and the height of the bar proportional to the frequency of the class. Since all the class widths are equal, the areas of the bars are also proportional to the class frequencies, and thus the bar chart may be considered to be a histogram. This method is slightly easier than the first method described but it is only applicable when all the class widths are equal.

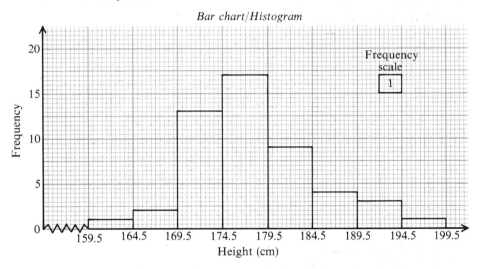

Bar chart/Histogram

It should be noted that the two histograms are identical apart from the scale on the vertical axis. The bar chart/histogram also has the advantage that it is easier to interpret, but it cannot be emphasised too much that the second method is limited in application.

Example 3

Construct a histogram to illustrate the grouped frequency distribution of the heights (measured to the nearest cm) of 50 men given in the following table.

Height (cm)	160–169	170–174	175–179	180–184	185–189	190–199
Frequency	3	13	17	9	4	4

Class	Frequency	Class boundaries		Class width	Frequency/class width
160–169	3	159.5	169.5	10	0.3
170–174	13	169.5	174.5	5	2.6
175–179	17	174.5	179.5	5	3.4
180–184	9	179.5	184.5	5	1.8
185–189	4	184.5	189.5	5	0.8
190–199	4	189.5	199.5	10	0.4

7

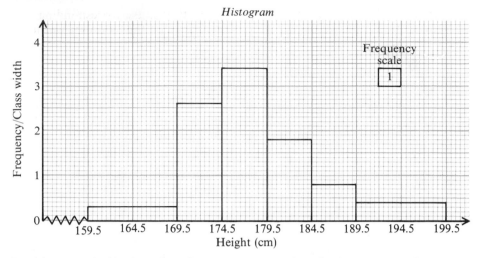

In this example, if a bar chart is constructed as described on page 7, the diagram obtained is *not* a histogram as the areas of the bars are not proportional to the frequencies of the classes because not all of the class widths are equal. Furthermore, the diagram is misleading as it gives a visual impression that the frequencies of the classes 160–169 and 190–199 are greater than is actually the case.

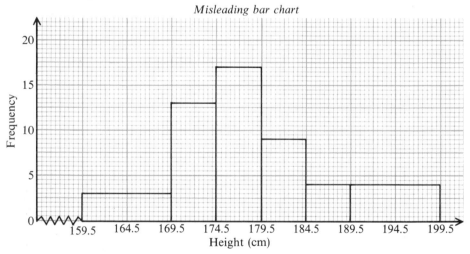

When the class intervals of a grouped frequency distribution are not all of equal width it is essential that the histogram to illustrate the distribution is constructed by the first method described.

Relative frequency histograms

A *relative frequency histogram* is a histogram in which the area of each rectangle is proportional to the relative frequency of each class. The *relative frequency* of a class is found by dividing the class frequency by the total frequency. A relative frequency histogram is identical to the corresponding histogram apart from the scale. It should be noted that the total area of the rectangles in a relative frequency histogram is 1.

Example 4

A grouped frequency distribution of the ages, in completed years, of 1000 children is given in the following table.

Age in completed years	5–6	7–10	11–15	16–17
Frequency	170	316	450	64

Draw a relative frequency histogram to represent this distribution.

A child with a nominal age of 5 has an age between 5 and 6, thus the lower class boundary of the first class is 5. Similarly the lower class boundaries of the other classes are 7, 11 and 16 respectively.

Class	Rel. freq.	Class boundaries		Class width	Rel. freq./Class width
5–6	0.170	5	7	2	0.085
7–10	0.316	7	11	4	0.079
11–15	0.450	11	16	5	0.090
16–17	0.064	16	18	2	0.032

The last column gives the heights of the rectangles.

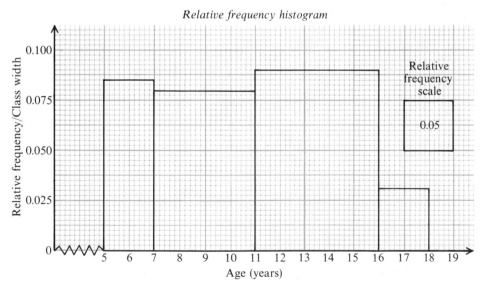

Relative frequency histogram

9

Important points to note about histograms:
1 the area of each rectangle is proportional to the frequency of the class;
2 there is no gap between successive rectangles;
3 the 'vertical axis' must start at zero;
4 the 'vertical axis' should be labelled frequency/class width;
5 the histogram should be drawn on graph paper.

Exercise 1.1

1 The numbers of goals scored in each of 46 football matches are recorded
below.
```
1 6 3 7 4 1 3 1 4 2 1 3 3 3 2 5 4 0 6 3 3 6 5
0 1 4 6 7 7 1 2 1 1 4 4 2 4 2 2 2 5 2 1 5 3 0
```
Find the frequency distribution of the number of goals scored per match
and construct a line diagram to illustrate these data.

2 Some married couples were asked to state how many children they had.
Their replies are recorded below.
```
1 0 0 2 0 3 1 1 0 0 0 4 3 3 2 0 1 1 1 0 2 0 2 2 3
0 1 5 3 1 0 1 1 3 3 4 0 0 0 2 2 0 1 2 2 3 4 1 1 2
```
Find the frequency distribution of the number of children per family and
construct a line diagram to illustrate these data.

3 The grouped frequency distribution of the ages (in completed years) of 1200
people living on an estate is given below.

Age	0–4	5–10	11–17	18–29	30–39	40–59	60–79
Frequency	128	144	161	240	284	181	62

Draw a histogram to represent this distribution.

4 The grouped frequency distribution of the masses (to the nearest kg) of
100 men is given below.

Mass	70–74	75–79	80–84	85–89	90–94	95–99	100–109
Frequency	16	21	28	19	8	6	2

Draw a relative frequency histogram to represent this distribution.

5 The grouped frequency distribution of the length of stay (measured to the
nearest minute) of cars at a short-term car-park is given below.

Length of stay	15–29	30–44	45–59	60–74	75–89	90–119
Frequency	56	63	87	123	67	22

Draw a histogram to represent this distribution.

6 The frequency distribution of the thicknesses (to the nearest 0.01 mm) of washers produced by a certain machine is given by the following table.

Thickness	0.96	0.97	0.98	0.99	1.00	1.01	1.02	1.03
Frequency	5	12	17	21	24	20	14	7

Draw a histogram to illustrate this distribution.

7 The grouped frequency distribution of the weight losses (to nearest 1 lb) of 76 women following a certain diet for sixteen weeks is given below.

Weight loss	0–5	6–9	10–13	14–17	18–21	22–25	26–35
Frequency	9	13	19	15	10	8	2

Draw a histogram to illustrate these data.

8 The frequency distribution of the ages (in completed years) of children in a comprehensive school is given in the following table.

Age	11	12	13	14	15	16	17	18
Frequency	182	175	190	181	168	83	64	7

Draw a histogram to illustrate this distribution.

1.2 Cumulative frequency distributions and diagrams

The *cumulative frequency* corresponding to a particular value x is the number of observations in the data set having values which are less than or equal to x.

Example 5

Find and illustrate the cumulative frequency distribution of the shoe sizes given in **Table 1**, which is reproduced below for ease of reference.

Table 1

Shoe size	6	$6\frac{1}{2}$	7	$7\frac{1}{2}$	8	$8\frac{1}{2}$	9	$9\frac{1}{2}$	10	$10\frac{1}{2}$	11
Frequency	2	3	5	7	10	17	18	13	11	8	6

The cumulative frequency of each shoe size is calculated from the frequency distribution given in **Table 1** by adding the frequency of the shoe size to the cumulative frequency of the previous shoe size. It should be checked that the cumulative frequency of the largest shoe size is equal to 100, the total number of men.

Thus, the cumulative frequency distribution of the shoe sizes given in **Table 1** is as shown in **Table 3** below.

Table 3

Shoe size	6	$6\frac{1}{2}$	7	$7\frac{1}{2}$	8	$8\frac{1}{2}$	9	$9\frac{1}{2}$	10	$10\frac{1}{2}$	11
Cumulative frequency	2	5	10	17	27	44	62	75	86	94	100

This distribution may be illustrated by the following cumulative frequency diagram. This is the graph of a step function since shoe size is a discrete variable.

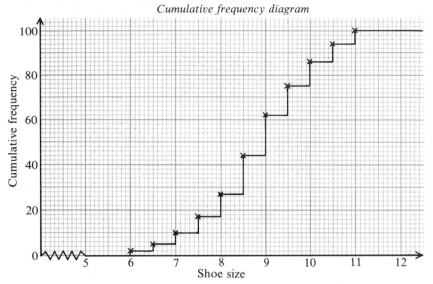

Cumulative frequency diagram

Grouped distributions

For data which have been grouped, the *cumulative frequency* of a class is the number of values of the variable which are less than the upper class boundary.

Example 6

Find and illustrate the cumulative frequency distribution of the heights given in **Table 2**, which is reproduced below for ease of reference.

Table 2

Height (cm)	160–164	165–169	170–174	175–179	180–184	185–189	190–194	195–199
Frequency	1	2	13	17	9	4	3	1

The cumulative frequency of each class is calculated from the grouped frequency distribution given in **Table 2** by adding the class frequency to the cumulative frequency of the previous class. It should be checked that the cumulative frequency of the last class is equal to the total frequency.

The cumulative frequency distribution of the heights given in **Table 2** is as shown in **Table 4** below. The abbreviation U.C.B. is used for upper class boundary.

Table 4

Class	160–164	165–169	170–174	175–179	180–184	185–189	190–194	195–199
U.C.B.	164.5	169.5	174.5	179.5	184.5	189.5	194.5	199.5
Cum. freq.	1	3	16	33	42	46	49	50

This distribution may be illustrated by means of a cumulative frequency polygon. The cumulative frequency of each class is plotted against its upper class boundary; then, for each class, it is assumed that the heights are uniformly distributed within the class and thus successive points are joined by straight lines.

The points plotted on the diagram below are $(159.5, 0)$, $(164.5, 1)$, $(169.5, 3)$, $(174.5, 16)$, $(179.5, 33)$, $(184.5, 42)$, $(189.5, 46)$, $(194.5, 49)$ and $(199.5, 50)$.

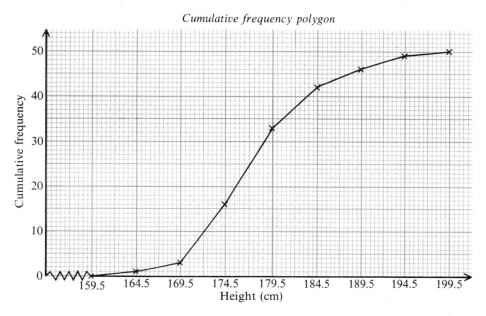

Cumulative frequency polygon

Example 7

The grouped frequency distribution of the marks obtained in an examination by 490 candidates is given in the table below.

Marks	0–19	20–29	30–39	40–49	50–59	60–69	70–79	80–89
Frequency	21	47	71	87	98	76	52	38

Find a cumulative frequency distribution and draw a cumulative frequency polygon.

As the mark of a candidate is a discrete variable, there is some difficulty in deciding the class boundaries. This difficulty is usually overcome by treating the candidate's mark as a continuous variable recorded by the examiner to the nearest whole number. If this is done the upper class boundaries are 19.5, 29.5, 39.5, etc. and the cumulative frequency distribution is as shown in the table below.

Class	0–19	20–29	30–39	40–49	50–59	60–69	70–79	80–89
U.C.B.	19.5	29.5	39.5	49.5	59.5	69.5	79.5	89.5
Cum. freq.	21	68	139	226	324	400	452	490

The points $(0, 0)$, $(19.5, 21)$, $(29.5, 68)$, $(39.5, 139)$, $(49.5, 226)$, $(59.5, 324)$, $(69.5, 400)$, $(79.5, 452)$, $(89.5, 490)$ are plotted and successive points joined with straight lines.

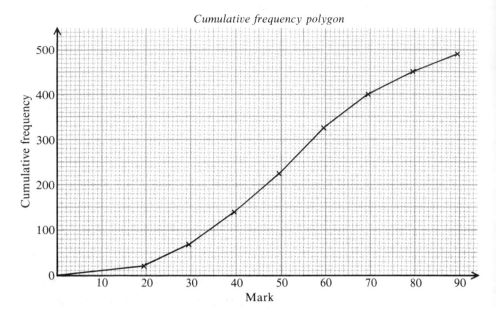

Cumulative frequency polygon

It should be noted that, although the above treatment is the most usual one, it is not unique. A candidate's mark could have been treated as a continuous variable recorded by the examiner to the nearest integer below the mark; in this case, the upper class boundaries would be 20, 30, 40, etc., giving a slightly different cumulative frequency polygon.

When the grouped frequency distribution relates to a sample taken from a larger set of data and the cumulative frequency diagram is to be used to estimate some aspect of the larger set of data, a smooth curve should be drawn through the points plotted on the cumulative frequency diagram. In this case the diagram is called a *cumulative frequency curve* or *ogive*. (See **Example 11** of Chapter 2, page 35.)

There are some important points to note about cumulative frequency diagrams:

1   if the data have been grouped, then the cumulative frequency must be plotted against the upper class boundary for each class and the plotted points joined by straight lines or by a smooth curve;

2   if the variable is discrete and the data ungrouped then the diagram is the graph of a step function;

3   the diagrams should be drawn on squared paper.

Exercise 1.2

1 The frequency distribution of the number of goals scored in 46 football matches is given in the table below.

Number of goals	0	1	2	3	4	5	6	7
Number of matches	4	8	8	8	7	4	4	3

Find the cumulative frequency distribution and draw a cumulative frequency diagram.

2 The frequency distribution of the number of children in 50 families is given in the table below.

Number of children	0	1	2	3	4	5
Number of families	15	13	10	8	3	1

Find the cumulative frequency distribution and draw a cumulative frequency diagram.

3 The grouped frequency distribution of the ages (in completed years) of 1200 people living on an estate is given below.

Age	0–4	5–10	11–17	18–29	30–39	40–59	60–79
Frequency	128	144	161	240	284	181	62

Construct a cumulative frequency distribution and draw a cumulative frequency diagram.

4 The grouped frequency distribution of the masses (to the nearest kg) of 100 men is given below.

Mass	70–74	75–79	80–84	85–89	90–94	95–99	100–109
Frequency	16	21	28	19	8	6	2

Construct a cumulative relative frequency distribution and draw a cumulative relative frequency diagram.

5 The grouped frequency distribution of the length of stay (to the nearest minute) of cars at a short-term car-park is given in the following table.

Length of stay	15–29	30–44	45–59	60–74	75–89	90–119
Frequency	56	63	87	123	67	22

Construct a cumulative frequency distribution and draw a cumulative frequency diagram.

6 The frequency distribution of the thicknesses (to the nearest 0.01 mm) of washers produced by a certain machine is given by the following table.

Thickness	0.96	0.97	0.98	0.99	1.00	1.01	1.02	1.03
Frequency	5	12	17	21	24	20	14	7

Construct a cumulative frequency distribution and draw a cumulative frequency polygon.

7 The grouped frequency distribution of the weight losses (to nearest 1 lb) of 76 women following a certain diet for sixteen weeks is given below.

Weight loss	0–5	6–9	10–13	14–17	18–21	22–25	26–35
Frequency	9	13	19	15	10	8	2

Construct a cumulative frequency distribution and draw a cumulative frequency polygon.

8 The frequency distribution of the ages (in completed years) of children in a comprehensive school is given by the following table.

Age	11	12	13	14	15	16	17	18
Frequency	182	175	190	181	168	83	64	7

Construct a cumulative frequency distribution and draw a cumulative frequency diagram.

1.3 Types of data

In statistics, data are collected in the form of measurements or observations of variables, e.g. heights of persons, temperatures, dress sizes, voting intentions in elections. Data on variables may be divided into four groups depending on the way the variable is measured or observed.

The highest level of measurement is that in which the ratio of any two values of a variable is independent of the unit of measurement; the variable is said to be measured on a *ratio scale*. The height of a person is an example of such a variable; if the ratio of the heights of two persons is 6:5 when the heights are measured in metres, then the ratio of the two heights is also 6:5 when the heights are measured in inches or any other units. A ratio scale has an arbitrary unit and a true zero point.

A variable such as temperature is not measured on a ratio scale, because the ratio of the two temperatures 100°C and 50°C measured on the Celsius scale is not the same as the ratio of the two corresponding temperatures 212°F and 122°F measured on the Fahrenheit scale. However the differences between 100°C and 50°C (212°F and 122°F) and between 80°C and 30°C (176°F and 86°F) which are equal when measured on the Celsius scale are also equal when measured on the Fahrenheit scale. Consider also historical time measured by a calendar; the ratio of the dates of two events is meaningless but the time interval between the dates is the same regardless of the calendar being used. Such variables are said to be measured on an *interval scale*. An interval scale has an arbitrary unit and an arbitrary zero point.

Other variables are measured on scales which are neither ratio nor interval; for example, a famous clothing manufacturer once made dresses with sizes denoted by the numbers 9, 10, 11, 12, 13, 14, 15, 16, 17, 18 in which an odd numbered size had the same bust, waist and hip measurements as the next even numbered size, but had a shorter length. On such a scale of measurement, the difference between size 13 and size 14 is not the same as the difference between size 14 and size 15, so the scale is not an interval scale. However, since it is possible to rank the dress sizes measured on this scale, this is an example of a variable being measured on an *ordinal scale*. Other methods of sizing dresses may be on interval scales.

When the observations of a variable cannot be ranked in a meaningful way, the variable is said to be *qualitative* or *categorical*. The voting intention of a person interviewed in an opinion poll is an example of a qualitative variable.

Data on such variables are called ratio, interval, ordinal and qualitative data, respectively.

Exercise 1.3

Classify each of the following variables (ratio, interval, ordinal, categorical).

1 The religious faith of a person.

2 The mass of a person.

3 The mark of a candidate in an examination.

4 The grade of a candidate in an examination.

5 The operational lifetime of a TV set.

6 The size of a man's shoe.

7 The position of a runner at the end of a race.

8 The intelligence quotient of a person.

9 The colour of a person's hair.

10 The time of a runner in a race.

Miscellaneous Exercise 1

1 A die was thrown 60 times and the following results were recorded:

```
1  1  4  3  5  2  3  1  6  6  6  1  4  3  2  3  5  4  6  1  2  5  3  3  4
5  6  2  3  4  5  6  2  4  5  6  2  1  6  3  5  6  3  2  2  1  6  6  5  2
3  3  5  1  3  1  1  3  2  3
```

Draw a line diagram to illustrate these data. Find the cumulative frequency distribution and draw an appropriate cumulative frequency diagram.

2 The grouped frequency distribution of the systolic blood pressures (to the nearest mm) of 141 men is tabulated below.

Blood pressure	105–114	115–119	120–124	125–129	130–134	135–139	140–144	145–154
Frequency	10	13	16	24	27	21	15	15

Draw a histogram to illustrate these data. Use the data to construct a cumulative frequency distribution and draw the cumulative frequency polygon.

3 The numbers of α-particles emitted from a radioactive source in 200 periods, each of 10 seconds duration, were as follows.

No. of particles	0	1	2	3	4	5	6	7	8	9	10
Frequency	2	7	17	29	36	35	30	21	13	7	3

Illustrate these data. Use the data to construct a cumulative frequency distribution and draw an appropriate cumulative frequency diagram.

4 The distribution of the cost, correct to the nearest £, of each of fifty summer courses is shown in the following table.

Cost (£)	40–49	50–54	55–59	60–64	65–69	70–79	80–89
Frequency	4	5	10	12	5	6	8

Draw a histogram to represent these data. Construct a cumulative frequency distribution and draw the cumulative frequency polygon.

5 The marks of 100 candidates in an examination are recorded below.

73 44 52 63 21 13 87 52 48 59 93 11 56 84 78 75 46 48
56 62 8 68 74 43 49 53 76 42 31 18 23 48 38 55 66 53
52 52 41 71 58 31 53 84 48 45 59 51 69 84 98 15 47 43
60 40 34 75 9 3 41 50 56 37 95 48 73 22 30 2 72 40
56 32 21 84 56 59 42 40 20 64 64 84 69 33 50 56 68 26
30 60 59 84 38 62 39 94 56 32

Select suitable classes and obtain a grouped frequency distribution. Draw a histogram to represent this distribution. Find a cumulative frequency distribution from the grouped frequency distribution and draw a cumulative frequency polygon to illustrate it.

Chapter 2

Summary measures

2.1 Measures of location and dispersion

Summary measures are numerical values which seek to quantify certain aspects of a data set. A *measure of location* is a value which is typical of the values in the data set and a *measure of dispersion* indicates the variability of the values within the data set. Such measures enable one to make quantitative comparisons between data sets, but before discussing these measures in detail, it is necessary to introduce a notation which may be unfamiliar to some readers.

Sigma notation

The *sigma* notation is so called because the Greek capital letter Σ (sigma) is used to mean 'the sum of all such terms as', e.g.

(i) $\displaystyle\sum_{r=1}^{5} r^2$ means the sum of all the terms obtained by successively putting r equal to 1, 2, 3, 4, 5 in the expression r^2,

$$\sum_{r=1}^{5} r^2 = 1^2 + 2^2 + 3^2 + 4^2 + 5^2.$$

(ii) $\displaystyle\sum_{i=1}^{3} x_i$ means the sum of all the terms obtained by successively putting i equal to 1, 2, 3 in the expression x_i,

$$\sum_{i=1}^{3} x_i = x_1 + x_2 + x_3.$$ (Note that x_1 denotes the first value of x, x_2 the second value and x_3 the third value.)

It should be noted that the letters r and i are 'dummy' variables, i.e. they are unimportant in themselves; the same results would have been obtained if other letters had been used. It follows that

$$\sum_{r=1}^{5} (2r + 1) = \sum_{i=1}^{5} (2i + 1) = \sum_{k=2}^{6} (2k - 1),$$

as may be verified by writing each in full.

The sigma notation is very useful and will be used extensively in this book. The following results (where k is a constant) are particularly useful:

1 $\displaystyle\sum_{i=1}^{n} k = nk;$

2 $\displaystyle\sum_{i=1}^{n} kx_i = k \sum_{i=1}^{n} x_i;$

3 $\displaystyle\sum_{i=1}^{n} (x_i + y_i) = \sum_{i=1}^{n} x_i + \sum_{i=1}^{n} y_i.$

Proofs:

1 $\displaystyle\sum_{i=1}^{n} k = (k + k + \ldots + k)$

$\displaystyle\sum_{i=1}^{n} k = nk$

2 $\displaystyle\sum_{i=1}^{n} kx_i = (kx_1 + kx_2 + \ldots + kx_n)$

$\displaystyle\sum_{i=1}^{n} kx_i = k(x_1 + x_2 + \ldots + x_n)$

$\displaystyle\sum_{i=1}^{n} kx_i = k \sum_{i=1}^{n} x_i$

3 $\displaystyle\sum_{i=1}^{n} (x_i + y_i) = [(x_1 + y_1) + (x_2 + y_2) + \ldots + (x_n + y_n)]$

$\displaystyle\sum_{i=1}^{n} (x_i + y_i) = [(x_1 + x_2 + \ldots + x_n) + (y_1 + y_2 + \ldots + y_n)]$

$\displaystyle\sum_{i=1}^{n} (x_i + y_i) = \sum_{i=1}^{n} x_i + \sum_{i=1}^{n} y_i$

Measures of location

A *measure of location* is a single value which acts as a representative of the values in the data set. One measure of location, which may be familiar to some readers, is a batting average in cricket (i.e. the number of runs scored divided by the number of completed innings); a batting average acts as a representative of all the scores obtained by a batsman. There are three common measures of location used in statistics—*mean*, *median* and *mode*.

Mean

The most commonly used measure of location is the arithmetic mean or, more simply, the mean. The *mean* $\bar{x}$ of the n values $x_1, x_2, x_3, \ldots, x_n$ is defined by

$$\bar{x} = \frac{\sum\limits_{i=1}^{n} x_i}{n}.$$

The mean of the 50 heights given in **Example 2** in Chapter 1 is calculated by dividing the sum, 8885 cm, of the heights by 50 giving an answer of 177.7 cm.

When the data are given in the form of a frequency distribution with the values x_i ($i = 1, 2, \ldots, k$) having frequencies f_i, then the *mean* $\bar{x}$ of this distribution is given by

$$\bar{x} = \frac{\sum\limits_{i=1}^{k} f_i x_i}{n} \qquad \text{where} \quad n = \sum\limits_{i=1}^{k} f_i.$$

When the data are given in the form of a grouped frequency distribution with k classes, f_i as the frequency of the ith class and x_i its mid-interval value. The *mean* $\bar{x}$ is then given by the same formula.

Example 1

Find the mean of the grouped frequency distribution of the heights given in **Table 2** in Chapter 1.

Class	160–164	165–169	170–174	175–179	180–184	185–189	190–194	195–199	Total
f_i	1	2	13	17	9	4	3	1	50
x_i	162	167	172	177	182	187	192	197	
$f_i x_i$	162	334	2236	3009	1638	748	576	197	8900

$$n = \Sigma f_i = 50 \text{ and } \Sigma f_i x_i = 8900 \qquad \therefore \bar{x} = \frac{8900}{50} = 178$$

Later in the chapter, an easier method of computation will be demonstrated. It should be noted that, although the mean of the ungrouped or raw data and the mean of the grouped data are approximately equal, they are not identical. The mean of the grouped data is used as an estimate for the mean of the raw data when the raw data are unavailable.

One reason why the mean is so often used as a measure of location is that its value depends upon all the values in the data set, but there are two cases where its use may be misleading:

1 when atypical values distort the mean, e.g. if the weights of four oarsmen and a cox are 90, 92, 95, 96 and 52 kg respectively, then the value of the mean, 85 kg, is less than four out of five of the weights;

2 when the data are ordinal so that the addition of values of the variable is inappropriate.

In these two cases, an alternative summary measure of location may be used — the median. The median is insensitive to extreme values in the data set and its calculation does not involve addition.

Median

If an odd number of values of a variable is arranged in order of magnitude, the *median* is that value of the variable for which there are equal numbers of values of the variable above and below it. The median is thus the value in the middle position, e.g. the median of 2, 3, 4, 4, 5, 6, 6, 7, 9 is 5.

If there is an even number of values, then the *median* is the arithmetic mean of the values in the two middle positions. Thus the median of 2, 3, 4, 5, 6, 9 is $\frac{(4 + 5)}{2} = 4.5$.

Example 2

Find the median of the shoe sizes given in **Example 1** of Chapter 1.

The median of the shoe sizes can be found easily from the cumulative frequency distribution given in **Table 3**, which is reproduced below for reference.

Table 3

Shoe size	6	$6\frac{1}{2}$	7	$7\frac{1}{2}$	8	$8\frac{1}{2}$	9	$9\frac{1}{2}$	10	$10\frac{1}{2}$	11
Cum. freq.	2	5	10	17	27	44	62	75	86	94	100

The values of the shoe size in the 50th and 51st positions are both 9; thus the median shoe size is 9.

The *median class* of a grouped frequency distribution is the class which contains the median.

Example 3

Find the median class of the grouped frequency distribution of the heights given in **Example 2** in Chapter 1.

It is easy to find the median class from the cumulative grouped frequency distribution given in **Table 4**, which is reproduced below for reference.

Table 4

Height	160–164	165–169	170–174	175–179	180–184	185–189	190–194	195–199
Cum. freq.	1	3	16	33	42	46	49	50

Since both the 25th and 26th heights are in the 175–179 class, the median class is 175–179.

The actual value of the median of the heights given in **Example 2** of Chapter 1 is more tedious to find (its value is 177 cm). Methods whereby estimates of the median may be calculated from the grouped frequency distribution will be shown later.

Since the median is centrally located and the mean tends to be centrally located within the data set, they are sometimes called *measures of central tendency*.

Mode

Another measure of location which is occasionally used is the *mode*, which is the value of the variable with the highest frequency.

Example 4

Find the mode of the shoe sizes in **Example 1** of Chapter 1.

The mode is easily found from the frequency distribution given in **Table 1**, which is reproduced below for ease of reference.

Table 1

Shoe size	6	$6\frac{1}{2}$	7	$7\frac{1}{2}$	8	$8\frac{1}{2}$	9	$9\frac{1}{2}$	10	$10\frac{1}{2}$	11
Frequency	2	3	5	7	10	17	18	13	11	8	6

Since shoe size 9 has the highest frequency, the mode is 9.

The *modal class* of a grouped frequency distribution is the class with the highest frequency.

Example 5

Find the modal class of the grouped frequency distribution of heights given in **Table 2** of Chapter 1, which is reproduced below for ease of reference.

Table 2

Class	160–164	165–169	170–174	175–179	180–184	185–189	190–194	195–199
Frequency	1	2	13	17	9	4	3	1

Since class 175–179 has the highest frequency, the modal class is 175–179.

The actual mode of the ungrouped data in **Example 2** of Chapter 1 is 172, which is not even in the modal class of the grouped data. In other examples the mode may not be unique. For these reasons, the mode is a much less useful measure than either the mean or the median.

Measures of dispersion

A measure of location is insufficient to summarise a set of data adequately. For example, suppose that the marks of a particular class in English vary from 21 to 83 with a mean mark of 50 and that the marks of the same class in Mathematics vary from 15 to 96 with a mean mark of 50; a mere comparison of the mean marks indicates no difference between the two sets, but clearly the marks in Mathematics are more widespread than the marks in English. A *measure of dispersion* is a number which indicates the variability of the values within a data set. A simple measure of dispersion is the range, which is the difference between the highest and lowest values of the variable in the set. In the example above, the range of the English marks is 62, whilst the range of the Mathematics marks is 81; the greater range of the Mathematics marks indicating that they vary to a greater extent than the English marks. The range is easily calculated, but it is of limited use. For theoretical reasons, which will become apparent later, the most widely used measures of dispersion are the standard deviation and the variance, both of which measure how close the values in the data set cluster around the mean. The square of the deviation of each value from the mean is calculated and the mean of these squares is called the variance; the positive square root of the variance is called the standard deviation.

Variance and standard deviation

The *variance* of the n values x_1, x_2, x_3, ..., x_n is defined by

$$\text{Variance} = \frac{\sum\limits_{i=1}^{n} (x_i - \bar{x})^2}{n} \quad \text{where} \quad \bar{x} = \frac{\sum\limits_{i=1}^{n} x_i}{n}.$$

This formula is not always very convenient for computational purposes, but it may be transformed as follows:

$$\sum_{i=1}^{n} (x_i - \bar{x})^2 = \sum_{i=1}^{n} (x_i^2 - 2x_i\bar{x} + \bar{x}^2)$$

$$= \sum_{i=1}^{n} x_i^2 - \sum_{i=1}^{n} 2x_i\bar{x} + \sum_{i=1}^{n} \bar{x}^2$$

since $\bar{x}$ is independent of i:

$$= \sum_{i=1}^{n} x_i^2 - 2\bar{x} \sum_{i=1}^{n} x_i + \bar{x}^2 \sum_{i=1}^{n} 1$$

since $\sum\limits_{i=1}^{n} x_i = n\bar{x}$:

$$= \sum_{i=1}^{n} x_i^2 - 2\bar{x}n\bar{x} + \bar{x}^2 n$$

$$\therefore \sum_{i=1}^{n} (x_i - \bar{x})^2 = \sum_{i=1}^{n} x_i^2 - n\bar{x}^2.$$

Thus an alternative formula for the variance is as follows:

$$\text{Variance} = \frac{\sum\limits_{i=1}^{n} x_i^2}{n} - \bar{x}^2.$$

The *standard deviation* of the set of values is defined as the positive square root of the variance.

Example 6

The frequency distribution of the number of goals scored in each of 50 football matches is given by the table below.

Number of goals	0	1	2	3	4	5	6	7
Number of matches	6	8	10	9	7	4	3	3

Calculate the mean and standard deviation of this distribution.

The following method shows how the mean and standard deviation may be calculated without the use of an electronic calculator.

x_i	f_i	$f_i x_i$	$f_i x_i^2$
0	6	0	0
1	8	8	8
2	10	20	40
3	9	27	81
4	7	28	112
5	4	20	100
6	3	18	108
7	3	21	147
Totals	50	142	596

$$\bar{x} = \frac{\sum x_i}{n}$$

$$\bar{x} = \frac{142}{50} = 2.84$$

$$\text{Variance} = \frac{\sum x_i^2}{n} - \bar{x}^2$$

$$\text{Variance} = \frac{596}{50} - 2.84^2 = 3.8544$$

Standard deviation $= \sqrt{3.8544} = 1.963$ (to 3 decimal places).

If the data are given in the form of a grouped frequency distribution with k classes and f_i is the frequency and x_i the mid-interval value of the ith class, then the variance of this distribution is defined by

$$\text{Variance} = \frac{\sum_{i=1}^{k} f_i(x_i - \bar{x})^2}{n} \quad \text{where} \quad n = \sum_{i=1}^{k} f_i \quad \text{and} \quad \bar{x} = \frac{\sum_{i=1}^{k} f_i x_i}{n}.$$

25

An alternative form, which is often more useful, is as follows:

$$\text{Variance} = \frac{\sum_{i=1}^{k} f_i x_i^2}{n} - \bar{x}^2.$$

When the raw data is unavailable, this variance is used as an approximation to the variance of the raw data; the approximation is often only moderately good.

Change of origin and scale

When the data are presented in the form of a grouped frequency distribution with classes of equal width, the computation of the mean and standard deviation may be simplified by means of a change of origin and scale, a process often called *coding*.

Let a be a mid-interval value of the distribution and b be the class width.

$$\text{Let} \qquad y_i = \frac{(x_i - a)}{b}$$

$$\text{then} \qquad x_i = by_i + a$$

$$\therefore \quad \bar{x} = \frac{\sum_{i=1}^{n} (by_i + a)}{n}$$

$$\bar{x} = \frac{\sum_{i=1}^{n} by_i + \sum_{i=1}^{n} a}{n}$$

$$\bar{x} = \frac{b \sum_{i=1}^{n} y_i + na}{n}$$

$$\bar{x} = b\bar{y} + a. \qquad (1)$$

Using this formula the value of $\bar{x}$ can be calculated from the value of $\bar{y}$.

Using (1)
$$x_i - \bar{x} = (by_i + a) - (b\bar{y} + a)$$
$$x_i - \bar{x} = b(y_i - \bar{y})$$

$$\therefore \quad \text{variance } (x) = \frac{\sum_{i=1}^{n} (x_i - \bar{x})^2}{n}$$

$$= \frac{\sum_{i=1}^{n} b^2 (y_i - \bar{y})^2}{n}$$

$$= \frac{b^2 \sum_{i=1}^{n} (y_i - \bar{y})^2}{n}$$

$$\text{variance } (x) = b^2 \text{ variance } (y) \qquad (2)$$

Using this formula, the variance of x can be calculated from the variance of y.

26

Coding may also be used to facilitate the calculation of the mean and variance of a frequency distribution of a discrete variable when the values of the variable are regularly spaced.

Example 7

Find the mean and standard deviation of the grouped frequency distribution of the heights given in **Table 2** of Chapter 1, reproduced below for reference.

Table 2

Height (cm)	160–164	165–169	170–174	175–179	180–184	185–189	190–194	195–199
Frequency	1	2	13	17	9	4	3	1

Let x_i be the mid-interval value of the ith class, $i = 1, 2, \ldots, 8$.

Let $y_i = \dfrac{(x_i - 177)}{5}$

	x_i	f_i	y_i	$f_i y_i$	$f_i y_i^2$
	162	1	-3	-3	9
	167	2	-2	-4	8
	172	13	-1	-13	13
	177	17	0	0	0
	182	9	1	9	9
	187	4	2	8	16
	192	3	3	9	27
	197	1	4	4	16
Totals		50		10	98

$$\bar{y} = \frac{\sum f_i y_i}{n} = \frac{10}{50} = 0.2$$

Using (1): $\bar{x} = 5\bar{y} + 177 = 178$

$$\text{variance } (y) = \frac{\sum f_i y_i^2}{n} - \bar{y}^2 = \frac{98}{50} - 0.2^2 = 1.92$$

Using (2): variance $(x) = 5^2$ variance $(y) = 48$

Mean $= 178$ and standard deviation $= \sqrt{48} = 6.93$ (2 d.p.).

For comparison purposes, it may be noted that the mean and the standard deviation of the raw data given in **Example 2** of Chapter 1 are 177.7 and 6.97 respectively.

Exercise 2.1

Find the mean and the standard deviation of each of the following distributions.

1

Number of goals	0	1	2	3	4	5	6	7	8
Frequency		11	25	16	35	21	19	12	10	1

2

No. of matches per box	45	46	47	48	49	50	51	52
Frequency	24	31	38	47	75	49	26	10

3

Age in completed years	20–29	30–39	40–49	50–59	60–69	70–79	80–89
Frequency	25	94	172	217	101	38	3

4

Weight (to nearest kg)	100–109	110–119	120–129	130–139	140–149	150–159
Frequency	4	23	64	72	32	5

5

Diameter (to nearest 0.01 cm)	3.25–3.27	3.28–3.30	3.31–3.33	3.34–3.36	3.37–3.39	3.40–3.42
Frequency	13	25	54	58	21	9

6 The frequency distribution of the thicknesses (to the nearest 0.01 mm) of washers produced by a certain machine is given by the following table.

Thickness	0.96	0.97	0.98	0.99	1.00	1.01	1.02	1.03
Frequency	5	12	17	21	24	20	14	7

7 The grouped frequency distribution of the weight losses (to nearest 1 lb) of 76 women following a certain diet for sixteen weeks is given below.

Weight loss	0–5	6–9	10–13	14–17	18–21	22–25	26–35
Frequency	9	13	19	15	10	8	2

8 The frequency distribution of the ages (in completed years) of children in a comprehensive school is given in the following table.

Age	11	12	13	14	15	16	17	18
Frequency	182	175	190	181	168	83	64	7

9 The grouped frequency distribution of the length of stay (in minutes) of cars at a short-term car-park is given in the following table.

Time	0–15	15–30	30–45	45–60	60–90	90–120
Frequency	54	84	102	110	36	24

10 The frequency distribution of the diameter (to nearest 0.01 mm) of bolts produced by a machine is given by the following table.

Diameter	25.37	25.38	25.39	25.40	25.41	25.42	25.43
Frequency	16	21	39	32	23	10	7

2.2 Percentiles

The median divides an ordered data set into two equal parts. This idea may be extended further; ideally, the *percentiles* of a data set should divide the ordered data set into one hundred equal parts. In practice this ideal cannot be achieved, but there are a number of ways in which suitable approximations may be calculated. Before describing one of these ways an alternative definition of the median is given below.

If there are n observations in the data set, then the median is associated with the cumulative frequency $\frac{1}{2}n$ in accordance with the following rules:

1 if $\frac{1}{2}n$ is an integer r, the median is the mean of the rth and $(r + 1)$th observations of the ordered set;

2 if $\frac{1}{2}n$ is not an integer but lies between the integers r and $(r + 1)$, then the median is the $(r + 1)$th observation in the ordered set.

The above definition is equivalent to the previous definition for the median, but it is in a form which may be more readily extended.

The kth percentile is a value associated with the cumulative frequency $\frac{kn}{100}$ in accordance with the following rules:

1 if $\frac{kn}{100}$ is an integer r, then the kth percentile is the mean of the rth and $(r + 1)$th observations in the ordered data set;

2 if $\frac{kn}{100}$ is not an integer but lies between the integers r and $(r + 1)$, then the kth percentile is the $(r + 1)$th observation in the ordered data set.

It should be noted that the 50th percentile is the median.

The above rules seem complicated but they are very easy to use in practice as the following example shows.

Example 8

Given the set of 15 observations 2, 3, 4, 5, 6, 7, 7, 7, 9, 11, 14, 14, 15, 16, 18, find

a the 25th percentile   **b** the 40th percentile
c the 50th percentile   **d** the 60th percentile.

a $\frac{kn}{100} = \frac{25 \times 15}{100} = 3.75$ (not an integer)

since 3.75 lies between 3 and 4, the 25th percentile is the 4th value

$$\therefore \quad \text{25th percentile} = 5$$

b $\frac{kn}{100} = \frac{40 \times 15}{100} = 6$ (an integer)

thus the 40th percentile is the mean of the 6th and 7th observations

$$\therefore \quad \text{40th percentile} = \frac{(7 + 7)}{2.} = 7$$

c $\frac{kn}{100} = \frac{50 \times 15}{100} = 7.5$ (not an integer)

since 7.5 lies between 7 and 8, the 50th percentile is the 8th value

$$\therefore \quad \text{50th percentile} = 7$$

d $\dfrac{kn}{100} = \dfrac{60 \times 15}{100} = 9$ (an integer)

thus the 60th percentile is the mean of the 9th and 10th values

$$\therefore \quad \text{60th percentile} = \dfrac{(9 + 11)}{2} = 10$$

Interquartile range

The 25th and 75th percentiles are called the *lower* and *upper quartiles* respectively. These two quartiles and the median, which is the 50th percentile, divide the ordered data set into four (approximately) equal parts. The difference between the upper and lower quartiles is called the *interquartile range*; the *semi-interquartile range* is half the interquartile range; both of these are used as measures of dispersion.

Example 9

Find the median and interquartile range of the shoe sizes given in **Example 1** of Chapter 1.

Two methods of answering this question are described below:

(i) By calculation from the cumulative frequency distribution

The cumulative frequency distribution is given in **Table 3** which is reproduced below for reference.

Table 3

Shoe size	6	$6\frac{1}{2}$	7	$7\frac{1}{2}$	8	$8\frac{1}{2}$	9	$9\frac{1}{2}$	10	$10\frac{1}{2}$	11
Cum. freq.	2	5	10	17	27	44	62	75	86	94	100

When $k = 50$, $\dfrac{kn}{100} = 50$ (an integer), thus the 50th percentile is the mean of the 50th and 51st values, both of which are 9.

$$\text{median} = 9$$

When $k = 25$, $\dfrac{kn}{100} = 25$ (an integer), thus the 25th percentile is the mean of the 25th and 26th values, both of which are 8.

$$\text{lower quartile} = 8$$

When $k = 75$, $\dfrac{kn}{100} = 75$ (an integer), thus the 75th percentile is the mean of the 75th and 76th values, which are $9\frac{1}{2}$ and 10 respectively.

$$\text{upper quartile} = 9.75$$
$$\text{interquartile range} = 9.75 - 8 = 1.75$$

(ii) Graphically from the cumulative frequency diagram

The cumulative frequency diagram was drawn in **Example 5** of Chapter 1 on page 12 and is reproduced below.

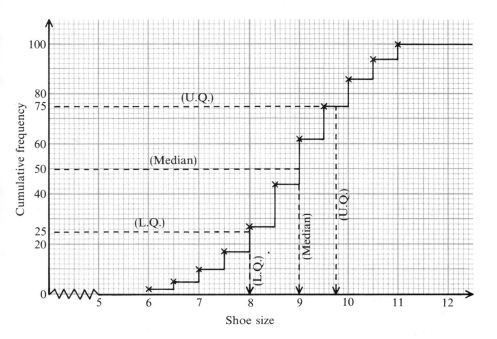

Having drawn the graph, it is easy to read off a value for any percentile. In this question the median corresponds to the cumulative frequency of 50 (50% of 100); the 'horizontal' line representing a cumulative frequency of 50 meets the graph at a point for which the shoe size is 9, thus the median is 9.

The 25th percentile corresponds to the cumulative frequency of 25 (25% of 100), giving a value of 8 for the lower quartile.

Since the 'horizontal' line representing the cumulative frequency of 75 meets the graph where a 'horizontal' step occurs, the cumulative frequency of 75 corresponds to an interval of values of shoe size; in this case the 75th percentile is taken as the mean of the extreme values of the interval, giving a value of $\frac{(9.5 + 10)}{2} = 9.75$ for the upper quartile.

It should be noted that other definitions for quartiles exist. These definitions and the one given above do not always give identical answers; the discrepancies are usually small and occur most frequently when the number of observations is small. This is not a matter for great concern; it simply reflects the imprecise nature of the task of dividing an ordered set of values into four equal parts. An advantage of the approach described above is its consistency with the procedures used for finding the quartiles of a grouped frequency distribution.

Example 10

Estimate the median and interquartile range of the 50 heights given in **Example 2** of Chapter 1 using the grouped frequency distribution given in **Table 2** of Chapter 1 on page 5.

Three methods for estimating the median and interquartile range from a grouped frequency distribution are described below.

(i) Graphically from a cumulative frequency polygon

The cumulative frequency polygon was drawn as in **Example 6** of Chapter 1 on page 13 and is reproduced below for reference. An estimated value m of the median may be found by reading from the cumulative frequency polygon the height corresponding to the cumulative frequency of 25 (50% of 50). Similarly estimated values l and u for the lower and upper quartiles may be found by reading the heights corresponding to cumulative frequencies of 12.5 (25% of 50) and 37.5 (75% of 50) respectively.

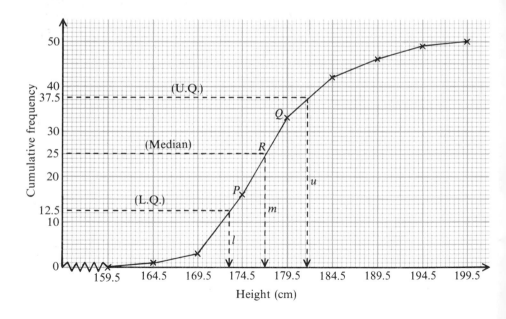

The values, to the nearest whole number, thus obtained are $m = 177$, $l = 173$ and $u = 182$. The estimated median = 177 and interquartile range = 9.

(ii) By calculation from the cumulative frequency polygon

The cumulative frequency distribution of the heights was given in **Table 4**, which is reproduced below for reference.

Table 4

Class	160–164	165–169	170–174	175–179	180–184	185–189	190–194	195–199
U.C.B.	164.5	169.5	174.5	179.5	184.5	189.5	194.5	199.5
Cum. freq.	1	3	16	33	42	46	49	50

The median is associated with the cumulative frequency 25, so it lies in the 4th class.

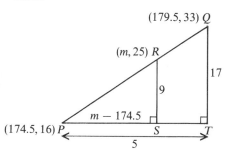

$$\frac{PS}{PT} = \frac{RS}{QT} \quad \text{(similar triangles)}$$

$$PS = \frac{RS}{QT} \times PT$$

Thus

$$m - 174.5 = \frac{(25 - 16)}{(33 - 16)} \times (179.5 - 174.5)$$

$$m - 174.5 = \frac{9}{17} \times 5$$

$$m = 177 \quad \text{(to the nearest whole number)}$$

The lower quartile is in the 3rd class, since it is associated with the cumulative frequency 12.5. Using a similar method to that shown above the values of l and u may be calculated.

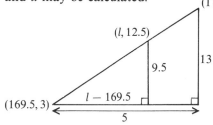

$$l - 169.5 = \frac{(12.5 - 3)}{(16 - 3)} \times (174.5 - 169.5)$$

$$l - 169.5 = \frac{9.5}{13} \times 5$$

$$l = 173 \quad \text{(to the nearest whole number)}$$

The upper quartile is in the 5th class, since it is associated with the cumulative frequency 37.5.

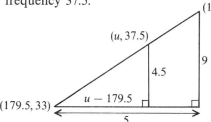

$$u - 179.5 = \frac{(37.5 - 33)}{(42 - 33)} \times (184.5 - 179.5)$$

$$u - 179.5 = \frac{4.5}{9} \times 5$$

$$u = 182$$

This method is called *linear interpolation*, since it has been assumed that, within each class, the cumulative frequency is linearly related to the height.

(iii) By calculation from a histogram

The histogram of the grouped frequency distribution was drawn in **Example 2** of Chapter 1 on page 6, and is reproduced below.

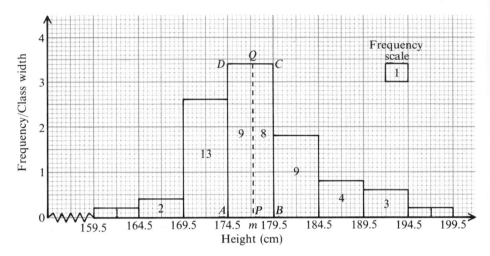

The vertical line PQ through m must divide the histogram into two parts which have equal areas. Since m is in the 4th class, the position of PQ must be such that the area of $APQD$ is equal to 9 making the total area to the left of PQ equal to 25.

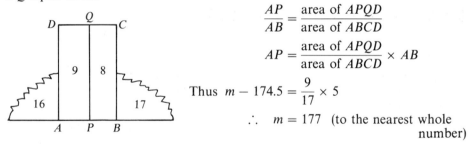

$$\frac{AP}{AB} = \frac{\text{area of } APQD}{\text{area of } ABCD}$$

$$AP = \frac{\text{area of } APQD}{\text{area of } ABCD} \times AB$$

Thus $m - 174.5 = \dfrac{9}{17} \times 5$

$$\therefore \quad m = 177 \quad \text{(to the nearest whole number)}$$

Similarly $l = 173$ and $u = 182$ (to the nearest whole number).

Exercise 2.2

Find the median and interquartile range of the following sets of numbers.

1 The set 1, 2, 4, 4, 5, 7, 8, 11, 15, 15.

2 The set 8, 8, 9, 12, 13, 15, 16, 17, 23, 25, 26, 27, 30, 33.

3 The set 34, 37, 38, 40, 41, 42, 45, 47, 48, 51, 52, 54, 55, 57, 58.

4 The set 12, 14, 15, 16, 18, 19, 20, 21, 23, 26, 30, 31.

5 The set 2, 7, 4, 6, 9, 5, 1, 3, 1, 3, 3, 7, 10, 9, 12, 15, 8, 4.

Find the median and interquartile range of each of the following distributions.

6

Number of goals	0	1	2	3	4	5	6	7	8
Frequency	11	25	16	35	21	19	12	10	1

7

No. of matches per box	45	46	47	48	49	50	51	52
Frequency	24	31	38	47	75	49	26	10

8

Age in completed years	20–29	30–39	40–49	50–59	60–69	70–79	80–89
Frequency	25	94	172	217	101	38	3

9

Mass (to nearest kg)	100–109	110–119	120–129	130–139	140–149	150–159
Frequency	4	23	64	72	32	5

10

Diameter (to nearest 0.01 cm)	3.25–3.27	3.28–3.30	3.31–3.33	3.34–3.36	3.37–3.39	3.40–3.42
Frequency	13	25	54	58	21	9

2.3 Further worked examples

Example 11

The grouped frequency distribution of the marks obtained in an examination by 490 candidates of the 12 585 candidates sitting the examination, is shown in the table below.

Marks	0–19	20–29	30–39	40–49	50–59	60–69	70–79	80–89	90–100
Frequency	21	47	71	87	98	76	52	38	0

Draw an appropriate cumulative frequency diagram to estimate the pass mark of the examination if it is known that 70.2 per cent of all candidates pass the examination.

The cumulative frequency distribution of the marks is given in the table below.

Class	0–19	20–29	30–39	40–49	50–59	60–69	70–79	80–89
U.C.B.	19.5	29.5	39.5	49.5	59.5	69.5	79.5	89.5
Cum. freq.	21	68	139	226	324	400	452	490

In this example the grouped frequency distribution relates to a sample of 490 marks taken from a larger set of 12 585 marks and the cumulative frequency diagram is to be used to estimate the pass mark of the larger set, so a smooth curve should be drawn through the points $(0, 0)$, $(19.5, 21)$, $(29.5, 68)$, $(39.5, 139)$, $(49.5, 226)$, $(59.5, 324)$, $(69.5, 400)$, $(79.5, 452)$, $(89.5, 490)$. Then, since just under 30% of the total candidature obtain a mark less than the pass mark, an estimate for the 30th percentile of all the marks is obtained by finding the mark corresponding to the cumulative frequency 147 (30% of 490) and the estimated pass mark is the largest whole number less than this estimate for the 30th percentile. This estimate will be a reasonable one if the marks of the 490 candidates are representative of the marks of the whole candidature.

The estimated pass mark is 40.

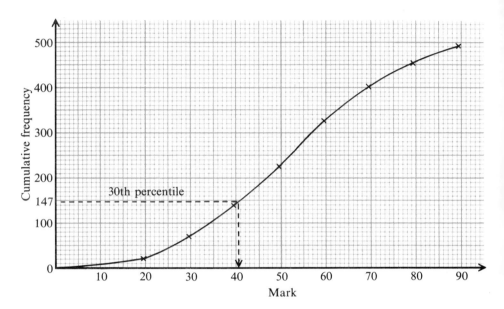

Example 12

a Find the mean and the variance of the set of numbers
   3, 4, 8, 9, 11.

b Deduce the mean and the variance of the sets of numbers
   (i) 5, 6, 10, 11, 13
   (ii) 5, 7, 15, 17, 21.

a $\sum x = 3 + 4 + 8 + 9 + 11 = 35$

$$\bar{x} = \frac{\sum x}{n}$$

$$\bar{x} = \frac{35}{5} = 7$$

$$\sum x^2 = 9 + 16 + 64 + 81 + 121 = 291$$

36

$$\text{variance } (x) = \frac{\sum x^2}{n} - \bar{x}^2$$

$$\text{variance } (x) = \frac{291}{5} - 7^2 = 9.2$$

b (i) Each number y in this set is given by $\qquad y = x + 2$
From equation (1) on page 26 $\qquad\qquad \bar{y} = \bar{x} + 2 \qquad \therefore \quad \bar{y} = 9$
From equation (2) on page 26 $\qquad$ variance $(y) = $ variance $(x) \qquad = 9.2$

(ii) Each number z in this set is given by $\qquad z = 2x - 1$
From equation (1) $\qquad\qquad\qquad\qquad \bar{z} = 2\bar{x} - 1 \qquad \therefore \quad \bar{z} = 13$
From equation (2) $\qquad\qquad$ variance $(z) = 2^2$ variance $(x) = 36.8$

Example 13

The mean and the standard deviation of the heights of a group of 30 women are 150 cm and 5 cm respectively; the mean and the standard deviation of the heights of a second group of 20 women are 155 cm and 6 cm respectively. Find the mean and the standard deviation of the heights of the combined group of 50 women.

For the first group:

$$\bar{x} = \frac{\sum x}{n} \qquad \Rightarrow 150 = \frac{\sum x}{30} \qquad \Rightarrow \sum x = 4500$$

$$\text{variance } (x) = \frac{\sum x^2}{n} - \bar{x}^2 \quad \Rightarrow \quad 5^2 = \frac{\sum x^2}{30} - 150^2 \quad \Rightarrow \quad \sum x^2 = 675750$$

For the second group:

$$\bar{x} = \frac{\sum x}{n} \qquad \Rightarrow 155 = \frac{\sum x}{20} \qquad \Rightarrow \sum x = 3100$$

$$\text{variance } (x) = \frac{\sum x^2}{n} - \bar{x}^2 \quad \Rightarrow \quad 6^2 = \frac{\sum x^2}{20} - 155^2 \quad \Rightarrow \quad \sum x^2 = 481220$$

For the combined group:

$$\sum x = 4500 + 3100 = 7600$$

$$\bar{x} = \frac{\sum x}{n} \quad \Rightarrow \quad \bar{x} = \frac{7600}{50} \quad \Rightarrow \quad \bar{x} = 152$$

$$\sum x^2 = 675750 + 481220 = 1156970$$

$$\text{variance } (x) = \frac{\sum x^2}{n} - \bar{x}^2 \quad \Rightarrow \quad \text{variance } (x) = \frac{1156970}{50} - 152^2 = 35.4$$

standard deviation $= \sqrt{35.4} = 5.950$ (3 d.p.)

Distributional shapes

Frequency distributions and grouped frequency distributions may be classified into broad categories depending upon their shapes.

For example:

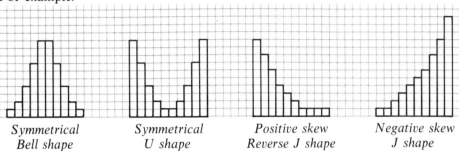

Symmetrical	*Symmetrical*	*Positive skew*	*Negative skew*
Bell shape	*U shape*	*Reverse J shape*	*J shape*

The mean and standard deviation are the summary measures which are most commonly used. They are preferred because they take into account all the recorded values of the variable. However, when the distribution is skew, the values of the mean and the standard deviation may be unduly affected by a few atypical values so, in this case, the median and interquartile range are generally used instead. If the distribution is U-shaped, both the mean and the median are atypical values of the distribution and the use of either as a single summary measure would be misleading.

Miscellaneous Exercise   2

1 a Find the mean and the variance of the set of numbers
   2, 5, 7, 7, 7, 9, 9, 10, 11, 13.

  b Deduce the mean and the variance of the sets of numbers
   (i) 3, 6, 8, 8, 8, 10, 10, 11, 12, 14
   (ii) 4, 10, 14, 14, 14, 18, 18, 20, 22, 26
   (iii) 5, 14, 20, 20, 20, 26, 26, 29, 32, 38.

2 a Find the mean and the variance of the set of numbers
   1, 1, 1, 2, 2, 2, 2, 3, 3, 3.

  b Deduce the mean and the variance of the sets of numbers
   (i) 3, 3, 3, 4, 4, 4, 4, 5, 5, 5
   (ii) 3, 3, 3, 6, 6, 6, 6, 9, 9, 9
   (iii) 3, 3, 3, 5, 5, 5, 5, 7, 7, 7.

3 The ages x_i ($i = 1, 2, 3, \ldots, 100$) of 100 persons were recorded in completed years and the following results were calculated: $\Sigma x_i = 3500$ and $\Sigma x_i^2 = 156725$. Calculate the mean and the standard deviation of these ages. Also calculate the mean and the standard deviation of the ages of these persons in 5 years time.

4 The mean and the standard deviation of the marks obtained in a test by the 17 boys in a class are 42 and 21 respectively; the mean and the standard deviation of the marks obtained in the test by the 13 girls in the class are 48 and 15 respectively. Find the mean and the standard deviation of the marks obtained by the 30 children in the class.

5 The mean and the standard deviation of the masses of 40 men are 80 kg and 3 kg respectively; the mean and the standard deviation of the masses of another group of 60 men are 85 kg and 5 kg respectively. Find the mean and the standard deviation of the masses of the combined group of 100 men.

6 The lengths of service, in completed years, of 8 employees working for a small firm are 1, 2, 2, 5, 7, 9, 14, 24. Calculate the mean and the standard deviation of the length of service of these employees.

If the longest serving employee retires and a new employee replaces him, calculate the mean and the standard deviation of the lengths of service of the new set of employees.

7 The mean and standard deviation of the marks obtained in a test by 20 boys in a class are 43 and 4 respectively; the mean and standard deviation of the marks obtained in the test by the 10 girls in the class are 46 and 5 respectively. Find the mean and the variance of the marks obtained by the 30 children in the class.

8 The grouped frequency distribution of the ages, in completed years, of 3500 school pupils is given in the following table.

Age	5–6	7–10	11–16	17–18
Number of pupils	613	1227	1564	96

Use these data to find a cumulative frequency distribution and draw a cumulative frequency diagram. Find estimates for the median and the interquartile range of these ages.

9 The distribution of the number of matches per box for 300 boxes is given by:

Number of matches per box	45	46	47	48	49	50	51	52
Number of boxes	24	31	38	47	75	49	26	10

Use these data to find the cumulative frequency distribution and draw a cumulative frequency diagram. Find the median and interquartile range of the number of matches per box.

10 The grouped frequency distribution of the size of cinemas (number of seats) is given by this table.

Size of cinema	100–199	200–399	400–599	600–799	800–999	1000–1499	1500–2500
Number of cinemas	2	41	73	95	62	18	9

Draw a histogram to illustrate the data. Estimate the median and interquartile range.

11 The numbers of alpha particles emitted from a radioactive source in 1000 periods of 5 seconds are recorded in the following table.

Number of particles	0	1	2	3	4	5	6	7
Frequency	32	195	226	282	179	43	31	12

Calculate the mean and standard deviation of this distribution.

12 The grouped frequency distribution of the marks obtained in a test by 263 candidates is given below.

Marks	0–19	20–39	40–59	60–79	80–100
Number of candidates	31	56	89	63	24

Use these data to find a cumulative frequency distribution and draw a cumulative frequency polygon. Estimate the median mark and the pass mark if 70 per cent of these candidates passed.

13 The grouped frequency distribution of the marks obtained in a test by 100 candidates is given below.

Marks	0–9	10–19	20–29	30–39	40–50
Number of candidates	7	18	35	25	15

Use these data to find a cumulative frequency distribution and draw a cumulative frequency diagram. Find estimates for the median and interquartile range of these marks. Also estimate the pass mark if 60 per cent of these candidates passed.

14 The following table shows the frequency distribution of the number of alpha particles emitted per second by a radioactive source.

Number of particles	0	1	2	3	4	5	6	7	8	9	10
Frequency	6	20	37	48	51	40	26	14	4	3	1

a Exhibit the distribution diagrammatically.

b Calculate the mean and the variance of the distribution.

c Find the median and the interquartile range.

15 The following table gives the cumulative frequency distribution of the masses, in kilograms, of a group of 200 youths.

U.C.B.	30	35	40	45	50	55	60	65	70	75	80	85	90	95
Cum. freq.	0	1	4	11	25	47	79	114	146	171	187	195	198	200

Draw a cumulative frequency polygon and estimate the median mass. Compile a grouped frequency distribution and deduce estimates for the mean and standard deviation of the masses of these youths.

16 The following table shows the grouped frequency distribution of the annual salaries, £x, of 120 employees of a firm.

£x	4000–4799	4800–5599	5600–6399	6400–7199
Frequency	20	44	22	17

£x	7200–8799	8800–10 399	10 400–11 999
Frequency	10	4	3

a Display the distribution as a histogram.

b Estimate the mean salary.

c Estimate the median salary. Give one reason why the median is a more appropriate measure of the average salary of these employees than the mean. Find an estimate of the interquartile range.

17 A factory employs a total of 200 men, of whom 120 have been at the factory for 10 years or longer. The grouped frequency distribution of the lengths of service, in completed years, of these 120 men is shown in the following table.

Length of service	10–14	15–19	20–24	25–29	30–39	40–49
Number of men	30	42	23	13	8	4

a Estimate the median and semi-interquartile range of the lengths of service of the 120 men.

b Estimate the median length of service of all 200 men employed at the factory.

c Given that the mean length of service of the 80 men with under 10 years service is 4.25 years, estimate the mean length of service of all 200 men employed at the factory. (*WJEC*)

18 The following table shows the distribution of the ages in years on their last birthdays of 120 students who registered for a particular degree course at a certain college on 1st October, 1974.

Age in years on 1/10/74	18	19	20	21	22	23
Number of students	50	32	22	8	5	3

a Estimate the mean age (in years and months) of these students on 1/10/74.

b Estimate to the nearest month in each case
(i) the median, and (ii) the semi-interquartile range, of the ages of these students on 1st July, 1977. (*WJEC*)

19 A business woman, who leaves home at 7.30 a.m. on each of five mornings each week to travel to work, keeps a record of the time, correct to the nearest minute, taken by her journey over a sixteen-week period. The results are shown in the following table.

Journey time in mins	74–76	77–79	80–82	83–85	86–88	89–91	92–94
Number of journeys	4	7	25	21	12	8	3

Use these data to construct a cumulative relative frequency table and draw on graph paper the cumulative relative frequency polygon.

Use your polygon to estimate for the journey times
 (i) a measure of average, clearly naming the measure you use
 (ii) the interquartile range
(iii) the percentage of journeys lasting longer than 86 minutes.

Assuming that these times are representative of the times taken for all her journeys, estimate the time which the businesswoman can state as her arrival time at the end of her journey so that there is no more than a 10 per cent chance that she will arrive late.

(You should smooth the cumulative frequency polygon.) *(JMB)*

20 A random sample of 1000 surnames is drawn from a local telephone directory. The distribution of the lengths of the names is as shown.

Number of letters in surname	3	4	5	6	7	8	9	10	11	12
Frequency			13	102	186	237	215	113	83	32	13	6

Calculate the sample mean and the sample standard deviation. Obtain the upper quartile.

Represent graphically the data in the table.

Give a reason why the sample of names obtained in this way may not be truly representative of the population of Great Britain. *(JMB)*

21 A man regularly runs round a fixed course. His times (to the nearest second) for 100 runs are distributed as shown in the following table.

Time	13m00s–13m29s	13m30s–13m59s	14m00s–14m29s	14m30s–14m59s	15m00s–15m29s
Frequency	2	3	16	23	25

Time	15m30s–15m59s	16m00s–16m29s	16m30s–16m59s	17m00s–17m29s
Frequency	18	7	4	2

State formulae for the mean and standard deviation of a frequency distribution, explaining the symbols used.

Prepare a table showing, for the above data, the mid-values of the class intervals and all the terms required for the summations that you need to calculate the mean and variance. Calculate the mean, variance and standard deviation, stating the units in each case.

Compile a cumulative frequency distribution and represent it graphically. Use your graph to estimate the median time. *(JMB)*

22 The grouped frequency distribution of the waiting times (to the nearest minute) of 200 patients attending a clinic is given in the following table.

Waiting time	0–2	3–7	8–12	13–17	18–22	23–27
Frequency	80	75	25	10	5	5

(i) Display the data as a histogram.

(ii) Estimate, showing your working, the mean and median waiting time.

23 The following table shows the frequency distribution of the masses, to the nearest kilogram, of a batch of 150 steel bars.

Mass	10–19	20–29	30–39	40–49	50–59	60–69	70–79	80–89
Frequency	4	18	28	56	25	15	3	1

Working with an origin at the midpoint of the fourth class interval and with units of 10 kilograms, prepare a table for the given data showing all terms in the summations that have to be calculated to obtain the mean and the standard deviation of the masses of the bars. Calculate these two quantities correct to one decimal place.

Compile a cumulative frequency distribution table and represent this distribution graphically by a cumulative frequency polygon. Using your graph, or otherwise, estimate the median mass of the bars.

Estimate from the cumulative frequency polygon, or otherwise, the smallest and the largest percentiles of the distribution which lie within two standard deviations of the mean. (*JMB*)

24 The following table shows the grouped frequency distribution of the ages, in completed years, of 2000 children.

Age in completed years	5 or 6	7–10	11–15	16 or 17
Number of children	300	500	1050	150

(i) Draw a histogram to represent this distribution.

(ii) Obtain, showing your method, estimates of the median and interquartile range of this distribution, giving your answers to the nearest month. (*JMB*)

25 The distribution of the times taken when a certain task was performed by each of a large number of people was such that its 20th percentile was 25 minutes, its 40th percentile was 50 minutes, its 60th percentile was 64 minutes and its 80th percentile was 74 minutes.

Use linear interpolation to estimate

(i) the median of the distribution

(ii) the upper quartile of the distribution

(iii) the percentage of persons who performed the task in 40 minutes or less. (*JMB*)

43

26 The degree of cloudiness of the sky may be measured on an eleven point scale, with the value 0 corresponding to a clear sky and the value 10 corresponding to a completely overcast sky. Observations of the degree of cloudiness were recorded at a particular meteorological station at noon each day during the month of June over ten consecutive years. The frequency distribution of the 300 observations was as follows.

Degree of cloudiness	0	1	2	3	4	5	6	7	8	9	10
Number of days	64	24	13	9	6	8	7	9	14	27	119

Illustrate the distribution graphically, and calculate the sample mean and variance.

Comment briefly on the shape of the distribution.

Discuss the usefulness (or otherwise) of the sample mean as a measure of location of such a distribution. (*JMB*)

Chapter 3

Probability

The theory of probability has been developed to allow quantitative statements to be made about situations in which there is an element of uncertainty. One type of situation of particular interest is called a random experiment. An experiment consists of a repeatable course of action and the observation of the result of that course of action. The result of the course of action is called the *outcome* of the experiment. If, whenever an experiment is carried out under identical conditions, the outcome is always the same then the experiment is said to be *deterministic*. An experiment whose outcome is not precisely predictable is called a *random experiment*. For example, a random experiment is carried out when an ordinary cubical die is rolled and the score is noted, or when a coin is tossed twice and the sequence of results is recorded.

Standard terms relating to random experiments are defined below; it is assumed that the reader is familiar with the notation and results of elementary set algebra.

3.1 Definitions

Sample spaces and events

The set S of all possible outcomes of a random experiment is called the *sample space* of the experiment. For example, the sample spaces for the two random experiments mentioned above are $\{1, 2, 3, 4, 5, 6\}$ and $\{HH, HT, TH, TT\}$.

An *event* is a characteristic associated with the outcome of a random experiment. Each of the elements of the sample space of the random experiment either has or does not have the characteristic; when the observed outcome has the characteristic, the event is said to have *occurred*. For example, when a coin is tossed twice in succession, a possible event is that 'one head is obtained'; this event occurs when the outcome of the random experiment is either HT or TH. In addition to a description of an event in words, the event can be represented by the subset of the sample space S which consists of all the elements of S having the characteristic defined by the verbal description of the event. Thus in the example above, the event 'one head is obtained' can be represented by the subset $\{HT, TH\}$ of S. Capital letters $A, B, C, \ldots$ will be used to denote events and their subset representations. So the letter A might be used to denote either the verbal description of the event 'one head is obtained' or its subset representation $\{HT, TH\}$.

A *sure* event is an event which must occur. A sure event can be represented by the sample space S.

An *impossible* event is an event which cannot occur. An impossible event can be represented by the empty set $\emptyset$.

Example 1

A fair coin is tossed twice in succession and the sequence of results is recorded. Write down the subset representations of the events described below.

a A is the event 'one tail is obtained'.

b B is the event 'a head is obtained on the first toss'.

c C is the event 'two tails are obtained'.

d D is the event 'at least one tail is obtained'.

Sample space $S = \{HH, HT, TH, TT\}$.

       a $A = \{HT, TH\}$.
       b $B = \{HH, HT\}$.
       c $C = \{TT\}$.
       d $D = \{HT, TH, TT\}$.

The event A', called the *complement* of A, is the event which occurs when A does not occur. In **Example 1**, $A' = \{HH, TT\}$.

The event $A \cup B$, called the *union* of A and B, is the event which occurs when A or B or both occur. In **Example 1**, $A \cup B = \{HH, HT, TH\}$.

The event $A \cap B$, called the *intersection* of A and B, is the event which occurs when both A and B occur. In **Example 1**, $A \cap B = \{HT\}$.

Two events which cannot occur simultaneously are said to be *mutually exclusive*. In **Example 1**, A and C are mutually exclusive since $A \cap C = \emptyset$.

Two events A and B, which are such that $A \cup B$ is a sure event, are said to be *exhaustive*. In **Example 1**, B and D are exhaustive since $B \cup D = S$.

Events $A_1, A_2, \ldots, A_n$ are said to be *mutually exclusive* if no two of the events can occur simultaneously.

Events $A_1, A_2, \ldots, A_n$ are said to be *exhaustive* if $A_1 \cup A_2 \cup \ldots \cup A_n$ is a sure event. In **Example 1**, A, B and C are exhaustive since $A \cup B \cup C = S$.

Classical definition of probability

The probability of an event is a measure of how likely it is that the event will occur. In many random experiments, the principle of symmetry may lead one to suppose that all outcomes in the sample space S are equally likely to occur, e.g. in the rolling of a fair die each of the possible scores 1, 2, 3, 4, 5, 6 is equally likely to occur. Confining attention to random experiments of this type, the following definition of probability is appropriate.

If a random experiment is such that its sample space S consists of equally likely outcomes, then the *probability* of an event A associated with the outcome of the experiment is given by

$$P(A) = \frac{n(A)}{n(S)},$$

where $n(A)$ and $n(S)$ are the numbers of elements in A and S respectively.

This is called the *classical definition of probability*.

It should be noted that the probability of a sure event is 1 and the probability of an impossible event is 0. Thus probability is measured on a scale between 0 and 1.

Example 2

A fair die is rolled once. Find the probability that the number uppermost is prime.

$S = \{1, 2, 3, 4, 5, 6\}$ $\qquad\qquad n(S) = 6$

Let A be the event which occurs when the number uppermost on the die is prime.

$A = \{2, 3, 5\}$ $\qquad\qquad n(A) = 3$ $\qquad \therefore\ P(A) = \frac{3}{6} = \frac{1}{2}$

Example 3

Two fair coins are tossed. Find the probability that at least one head is tossed.

$S = \{HH, HT, TH, TT\}$ $\qquad\qquad n(S) = 4$

Let A be the event which occurs when at least one head is tossed.

$A = \{HH, HT, TH\}$ $\qquad\qquad n(A) = 3$ $\qquad \therefore\ P(A) = \frac{3}{4}$

Exercise 3.1

In each of the following questions, list a sample space consisting of equally likely outcomes and hence find the required probabilities.

1 A fair coin is tossed 3 times. Find the probability that
   a 2 heads and 1 tail are tossed
   b 3 heads are tossed.

2 There are 5 counters in a bag, each marked with a different number from the set $\{1, 2, 3, 4, 5\}$. Two counters are drawn from the bag. Find the probability that the total score on the counters is
   a greater than 7
   b a prime number
   c an odd number.

3 There are 3 red balls and 1 white ball in a bag. Two balls are drawn from the bag. Find the probability that
   a both are red
   b the first is red and the second is white.

4 A fair coin is tossed 4 times. Find the probability that
   a 2 heads and 2 tails are tossed
   b a run of at least 3 successive tails is tossed.

5 A fair die is rolled twice. Find the probability that the total score is
   a exactly 7
   b at least 4
   c more than 12.

The method of listing the members of the sample space and deducing the required probabilities is often long and tedious; to reduce the amount of work it is necessary to establish some rules of probability.

3.2 Rules of probability

In the following, the results are established on the left of the page using formal set notation and on the right in an informal manner using Venn diagrams in which $n(S) = n$, $n(A) = a$, $n(B) = b$ and $n(A \cap B) = x$.

Rule 1 $P(A') = 1 - P(A)$.

$$P(A') = \frac{n(A')}{n(S)}$$

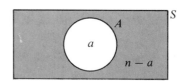

Since A', the complement of A, consists of all the elements of S which are not in A,

$$P(A') = \frac{n(S) - n(A)}{n(S)} \quad \text{or} \quad \frac{n - a}{n}$$

$$= \frac{n(S)}{n(S)} - \frac{n(A)}{n(S)} \quad \text{or} \quad \frac{n}{n} - \frac{a}{n}$$

$$= 1 - P(A).$$

Rule 2 $P(A \cup B) = P(A) + P(B) - P(A \cap B)$.

$$P(A \cup B) = \frac{n(A \cup B)}{n(S)} \quad \text{or} \quad \frac{(a - x) + x + (b - x)}{n}$$

The set $A \cup B$ consists of the elements of S which are in at least one of A and B; when $n(A)$ is added to $n(B)$ the number of elements in $A \cap B$ is counted twice, thus $n(A \cup B) = n(A) + n(B) - n(A \cap B)$.

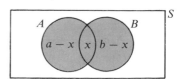

Hence $\quad \mathrm{P}(A \cup B) = \dfrac{n(A) + n(B) - n(A \cap B)}{n(S)} \qquad$ or $\qquad \dfrac{a + b - x}{n}$

$\qquad\qquad\quad = \dfrac{n(A)}{n(S)} + \dfrac{n(B)}{n(S)} - \dfrac{n(A \cap B)}{n(S)} \qquad$ or $\qquad \dfrac{a}{n} + \dfrac{b}{n} - \dfrac{x}{n}$

$\qquad\qquad\quad = \mathrm{P}(A) + \mathrm{P}(B) - \mathrm{P}(A \cap B).$

Rule 3 If A and B are mutually exclusive, then $\mathrm{P}(A \cup B) = \mathrm{P}(A) + \mathrm{P}(B)$.

If A and B are mutually exclusive, then A
and B have no element in common so that
$n(A \cap B) = 0$ and therefore $\mathrm{P}(A \cap B) = 0$.
Substituting in Rule 2,

$\qquad \mathrm{P}(A \cup B) = \mathrm{P}(A) + \mathrm{P}(B).$

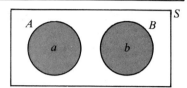

Rule 4 $\mathrm{P}(A \cap B') = \mathrm{P}(A) - \mathrm{P}(A \cap B)$.

Since B and B' are exhaustive, the event A
must occur either in combination with B or
in combination with B'. Thus

$\qquad A = (A \cap B) \cup (A \cap B'),$

and since the events $(A \cap B)$ and $(A \cap B')$ are
mutually exclusive, it follows from Rule 3,

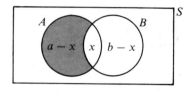

$\qquad \mathrm{P}(A) = \mathrm{P}(A \cap B) + \mathrm{P}(A \cap B') \quad$ or $\quad \mathrm{P}(A \cap B') = \dfrac{a - x}{n} = \dfrac{a}{n} - \dfrac{x}{n}.$

Rearranging this equation, the probability that of the two events A, B only A
will occur is

$\qquad \mathrm{P}(A \cap B') = \mathrm{P}(A) - \mathrm{P}(A \cap B).$

Example 4

Given that A and B are events such that
$\mathrm{P}(A) = 0.4, \qquad\qquad \mathrm{P}(B) = 0.3, \qquad\qquad \mathrm{P}(A \cap B) = 0.2,$
find
    a $\mathrm{P}(A')$
    b $\mathrm{P}(A \cup B)$
    c $\mathrm{P}(A' \cap B')$
    d the probability that only one of A and B occurs.

a Using Rule 1, $\qquad\qquad \mathrm{P}(A') = 1 - \mathrm{P}(A)$
$\qquad\qquad\qquad\qquad\qquad\quad = 1 - 0.4$
$\qquad\qquad\qquad\qquad\qquad\quad = 0.6$

b Using Rule 2, $\qquad \mathrm{P}(A \cup B) = \mathrm{P}(A) + \mathrm{P}(B) - \mathrm{P}(A \cap B)$
$\qquad\qquad\qquad\qquad\qquad\quad = 0.4 + 0.3 - 0.2$
$\qquad\qquad\qquad\qquad\qquad\quad = 0.5$

c Since $A' \cap B' = (A \cup B)'$,

using Rule 1, $\quad P(A' \cap B') = 1 - P(A \cup B)$
$$= 1 - 0.5$$
$$= 0.5$$

d Using Rule 4, $\quad P(A \cap B') = P(A) - P(A \cap B)$
$$= 0.4 - 0.2$$
$$= 0.2$$

Using Rule 4 with A and B interchanged,
$$P(A' \cap B) = P(B) - P(A \cap B)$$
$$= 0.3 - 0.2$$
$$= 0.1$$

Since $(A \cap B')$ and $(A' \cap B)$ are mutually exclusive,
$$P(\text{only one of } A, B) = 0.2 + 0.1$$
$$= 0.3$$

Exercise 3.2

1 Given that A and B are events such that
$$P(A) = 0.7, \qquad P(B) = 0.2, \qquad P(A \cap B) = 0.1,$$
find **a** $P(A')$ **b** $P(A \cup B)$.

2 Given that A and B are mutually exclusive events such that
$$P(A) = 0.6, \qquad P(B) = 0.3,$$
find **a** $P(A \cup B)$ **b** $P(A' \cap B')$.

3 Given that A and B are two events such that
$$P(A) = 0.5, \qquad P(B) = 0.3, \qquad P(A \cup B) = 0.6,$$
find **a** $P(A \cap B)$ **b** $P(A' \cap B)$.

4 Given that A and B are events such that
$$P(A') = 0.4, \qquad P(B') = 0.3, \qquad P(A \cap B) = 0.5,$$
find **a** $P(A \cup B)$
b the probability that only one of A, B occurs.

5 Given that A and B are two events such that
$$P(A) = 0.2, \qquad P(B) = 0.6, \qquad P(A' \cap B') = 0.2,$$
show that A and B are mutually exclusive.

3.3 Conditional probability and independent events

The conditional probability that B will occur given that A has occurred is denoted by $P(B|A)$.

> **Rule 5** $\quad P(B|A) = \dfrac{P(A \cap B)}{P(A)}.$

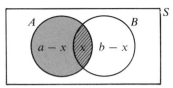

Given that A has occurred, the sample space now consists of the elements of A only, and the event B will occur only when the outcome is an element of $A \cap B$. Thus,

$$P(B|A) = \frac{n(A \cap B)}{n(A)} \qquad \text{or} \qquad \frac{x}{a}$$

$$= \frac{\dfrac{n(A \cap B)}{n(S)}}{\dfrac{n(A)}{n(S)}} \qquad \text{or} \qquad \frac{\dfrac{x}{n}}{\dfrac{a}{n}}$$

$$= \frac{P(A \cap B)}{P(A)}.$$

This result is often used in the form: $\quad P(A \cap B) = P(A) . P(B|A).$

Independent events

An event B is said to be independent of the event A if, and only if,

$$P(B|A) = P(B),$$

i.e. if, and only if, the probability of B is unaffected by the knowledge that A has occurred.

It is easy to show that the event A is independent of the event B if, and only if, the event B is independent of the event A. For, using Rule 5,

$$P(A|B) = \frac{P(A \cap B)}{P(B)},$$

and using the second version of Rule 5,

$$P(A|B) = \frac{P(A) . P(B|A)}{P(B)}.$$

It follows that $P(A|B) = P(A)$ if, and only if, $P(B|A) = P(B)$. Independence is therefore a mutual property of two events; in other words, events A and B are either independent of each other or not.

Since $\qquad P(B|A) = \dfrac{P(A \cap B)}{P(A)},$

it follows that $P(B|A) = P(B)$ if and only if

$$P(A \cap B) = P(A) . P(B).$$

This leads to an alternative definition of the independence of two events which is easier to use and more readily extended to include three or more events.

> **Rule 6** $\quad A$ and B are independent if, and only if, $P(A \cap B) = P(A) . P(B).$

Three or more events

Three events A, B, C are said to be *pairwise independent* if, and only if,

$$P(A \cap B) = P(A).P(B), \quad P(B \cap C) = P(B).P(C) \quad \text{and} \quad P(C \cap A) = P(C).P(A).$$

They are said to be *totally independent*, or more simply *independent* if, and only if, in addition,

$$P(A \cap B \cap C) = P(A).P(B).P(C).$$

It should be noted that pairwise independence does not necessarily imply total independence as may be seen in **Example 10** on page 57.

For the independence of more than three events the rule is more complicated and it will not be stated formally here. It is necessary that, for any combination of some or all of the events, the probability of their intersection is equal to the product of the probabilities of the individual events.

Other rules of probability may be deduced from the rules stated above; in some questions it may be useful to refer to a Venn diagram.

Example 5

Two events A and B are such that $P(A) = \frac{1}{3}$, $P(B) = \frac{1}{4}$ and $P(A \cup B) = \frac{1}{2}$. Find $P(A \cap B)$ and deduce that A and B are independent.

$$P(A \cup B) = P(A) + P(B) - P(A \cap B)$$

$$\frac{1}{2} = \frac{1}{3} + \frac{1}{4} - P(A \cap B)$$

$$\therefore \quad P(A \cap B) = \frac{1}{3} + \frac{1}{4} - \frac{1}{2}$$

$$P(A \cap B) = \frac{1}{12}$$

$$\text{Also} \quad P(A).P(B) = \frac{1}{3} \times \frac{1}{4} = \frac{1}{12}$$

$$\therefore \quad P(A \cap B) = P(A).P(B)$$

$\therefore$ A and B are independent.

Exercise 3.3

1 Given that A and B are two events such that

$$P(A) = 0.4, \qquad P(B) = 0.3, \qquad P(A \cup B) = 0.5,$$

find   **a** $P(A \cap B)$ **b** $P(B|A)$.

2 Given that A and B are two independent events such that

$$P(A) = 0.5, \qquad P(B) = 0.4,$$

find   **a** $P(A \cap B)$ **b** $P(A \cup B)$.

3 Given that A and B are two events such that

$$P(A) = 0.3, \qquad P(B \mid A) = 0.2, \qquad P(A \cup B) = 0.5,$$

    find   **a** $P(A \cap B)$ **b** $P(A \mid B)$.

4 Given that A and B are two events such that

$$P(A) = 0.6, \qquad P(B') = 0.7, \qquad P(A \cup B) = 0.72,$$

    show that A and B are independent.

5 Given that A, B, C are three independent events such that

$$P(A) = 0.4, \qquad P(B) = 0.5, \qquad P(C) = 0.2,$$

    find   **a** $P(A \cap B \cap C)$ **b** $P(A \cup B \cup C)$.

3.4 Alternative definition of probability

The classical definition of probability is not entirely satisfactory for there are many random experiments in which the outcomes are not equally likely. An alternative definition is needed in such cases.

If in n trials of a random experiment an event A occurs r times, then $\frac{r}{n}$, the proportion of the number of trials in which A occurs, is called the *relative frequency* of A. The limiting value of this relative frequency as $n \to \infty$ is defined as the *probability* of the event A.

This definition is based on the hypothesis that the limit $P(A)$ exists. The truth of this hypothesis is impossible to prove, but it is intuitively plausible and experimental evidence confirms that as the number of trials increases, the value of the relative frequency tends to stabilise around a fixed number. With this definition of probability the use of the rules of probability may be justified in a manner similar to that shown previously.

One of the greatest difficulties in applying the theory of probability to real-life situations is the assignment of realistic probabilities. The relative frequency approach provides a practical method whereby probabilities may be assigned to a wide range of events within the framework of a mathematical model. However, there is another type of event for which it is not possible to assign a probability by either the classical or relative frequency approaches. For example, the probability that man will land on Mars within the next twenty years cannot be assigned by the classical approach as the outcomes are not equally likely by symmetry, and since there is no past information relating to the event a probability cannot be assigned by the relative frequency approach. Nonetheless, a probability may be assigned to the event subjectively, i.e. based on the knowledge and experience of the person making the assignment but independent of direct experimental evidence. This subjective probability may be modified in the light of later events.

In view of the limitations of the classical and relative frequency approaches, more advanced courses adopt an axiomatic approach to establish the rules of probability in a rigorous manner. The axiomatic approach allows the rules of probability to be applied to probabilities however they have been assigned.

Further worked examples

Example 6

A company operates two machines A and B which function independently. The probability that both A and B break down within a month of being serviced is $\dfrac{1}{10}$ and the probability neither machine breaks down within the period is $\dfrac{9}{20}$. If A is more likely to break down than B, find the probability that A breaks down within a month of being serviced.

Let A be the event that machine A will break down within the stated period.
Let B be the event that machine B will break down within the stated period.

Let $P(A) = a$ and $P(B) = b$.
Since A and B are independent: $\quad P(A \cap B) = P(A) . P(B)$

$$\frac{1}{10} = ab \qquad (1)$$

Since $A \cup B$ consists of all the elements of S which are not in $A' \cap B'$:

$$P(A \cup B) = 1 - P(A' \cap B')$$

$$P(A \cup B) = 1 - \frac{9}{20} = \frac{11}{20}$$

but $\qquad\qquad\qquad P(A \cup B) = P(A) + P(B) - P(A \cap B)$

$$\frac{11}{20} = a + b - \frac{1}{10}$$

$$a + b = \frac{13}{20} \qquad (2)$$

Substituting for b from (2) in (1):

$$a\left(\frac{13}{20} - a\right) = \frac{1}{10}$$

$$\frac{13}{20}a - a^2 = \frac{1}{10}$$

Multiply by 20: $\qquad\qquad 13a - 20a^2 = 2$

$$20a^2 - 13a + 2 = 0$$

$$(5a - 2)(4a - 1) = 0$$

$$a = \frac{2}{5} \quad \text{or} \quad a = \frac{1}{4}$$

Substituting in (2): $\qquad\qquad b = \frac{1}{4} \quad \text{or} \quad b = \frac{2}{5}$

Since $P(A) > P(B)$: $\qquad\qquad P(A) = \frac{2}{5}, \ P(B) = \frac{1}{4}$

Example 7

A fair coin is tossed three times. Find the probability that the third toss is a head in each of the following cases.

a Given no further information.

b Given that exactly one head has been tossed.

c Given that at least one head has been tossed.

Sample space $S = \{HHH, HHT, HTH, THH, HTT, THT, TTH, TTT\}$

It is assumed that the coin is tossed in such a way that all the elements of S are equally likely to occur.

Let E_1 be the event that the third toss is a head.

Let E_2 be the event that exactly one head is tossed.

Let E_3 be the event that at least one head is tossed.

a
$$E_1 = \{HHH, HTH, THH, TTH\}$$

$$P(E_1) = \frac{n(E_1)}{n(S)} = \frac{4}{8} = \frac{1}{2}$$

b
$$E_2 = \{HTT, THT, TTH\}$$

$$P(E_2) = \frac{n(E_2)}{n(S)} = \frac{3}{8}$$

$$E_1 \cap E_2 = \{TTH\}$$

$$P(E_1 \cap E_2) = \frac{n(E_1 \cap E_2)}{n(S)} = \frac{1}{8}$$

$$P(E_1 | E_2) = \frac{P(E_1 \cap E_2)}{P(E_2)} = \frac{\frac{1}{8}}{\frac{3}{8}} = \frac{1}{3}$$

c This part may be answered using a method similar to that used in part **b** or by the following method, which is also applicable to part **b**.

Given that E_3 has happened means that the outcome must have been one of the elements of the new sample space S_1, where

$$S_1 = \{HHH, HHT, HTH, THH, HTT, THT, TTH\}.$$

Now
$$E_1 = \{HHH, HTH, THH, TTH\}$$

$$P(E_1 | E_3) = \frac{n(E_1)}{n(S_1)} = \frac{4}{7}$$

Example 8

A bag contains five balls which are numbered 1, 2, 3, 4, 5. Two balls are drawn at random, without replacement, from the bag. Find the probability that the sum of the two numbers drawn is

a a prime number **b** divisible by 3.

In this example the phrase 'drawn at random' means that each pair of balls has the same chance of being drawn. Furthermore, for the events described, the order in which the balls are drawn is not important, so the sample space S may be considered to consist of the following equally likely unordered pairs.

$$S = \{(1,2), (1,3), (1,4), (1,5), (2,3), (2,4), (2,5), (3,4), (3,5), (4,5)\}$$

a Let A be the event that the sum of the two numbers drawn is prime.

$$A = \{(1,2), (1,4), (2,3), (2,5), (3,4)\}$$

$$P(A) = \frac{n(A)}{n(S)} = \frac{5}{10} = \frac{1}{2}$$

b Let B be the event that the sum of the two numbers drawn is divisible by 3.

$$B = \{(1,2), (1,5), (2,4), (4,5)\}$$

$$P(B) = \frac{n(B)}{n(S)} = \frac{4}{10} = \frac{2}{5}$$

Example 9

Two persons are to be chosen at random from a group consisting of 6 men and 4 women. Calculate the probability that the two persons selected are

a both men **b** of different sex **c** of the same sex.

a Let M_1 be the event that the first person chosen is male.

Let M_2 be the event that the second person chosen is male.

Probability required $= P(M_1 \cap M_2)$

$$= P(M_1) . P(M_2 | M_1)$$

$$= \frac{6}{10} \times \frac{5}{9}$$

$$= \frac{1}{3}$$

b Let F_1 be the event that the first person chosen is female.

Let F_2 be the event that the second person chosen is female.

Probability required $= P(M_1 \cap F_2 \text{ or } F_1 \cap M_2)$

Since the compound events $M_1 \cap F_2$ and $F_1 \cap M_2$ are mutually exclusive

Probability required $= P(M_1 \cap F_2) + P(F_1 \cap M_2)$

$$= P(M_1) . P(F_2 | M_1) + P(F_1) . P(M_2 | F_1)$$

$$= \frac{6}{10} \times \frac{4}{9} + \frac{4}{10} \times \frac{6}{9}$$

$$= \frac{8}{15}$$

c Probability required $= P(M_1 \cap M_2 \text{ or } F_1 \cap F_2)$

Since the compound events $M_1 \cap M_2$ and $F_1 \cap F_2$ are mutually exclusive

Probability required $= P(M_1 \cap M_2) + P(F_1 \cap F_2)$

$$= \frac{6}{10} \times \frac{5}{9} + \frac{4}{10} \times \frac{3}{9}$$

$$= \frac{7}{15}$$

Example 10

The events A, B, C are pairwise independent and
$$P(A) = 0.5, \quad P(B) = 0.3, \quad P(C) = 0.2, \quad P(A \cup B \cup C) = 0.71.$$
Show that A, B, C are not totally independent.

Since A, B, C are pairwise independent

$$P(A \cap B) = 0.5 \times 0.3 = 0.15$$
$$P(B \cap C) = 0.3 \times 0.2 = 0.06$$
$$P(C \cap A) = 0.2 \times 0.5 = 0.1.$$

Let $P(A \cap B \cap C) = x$ and in each region of the Venn diagram enter the corresponding probability in terms of x.

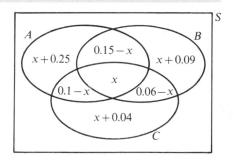

The sum of the probabilities entered in the diagram is equal to $P(A \cup B \cup C)$.

$$(x + 0.25) + (0.15 - x) + (x + 0.09) + (0.1 - x) + x + (0.06 - x) + (x + 0.04) = 0.71$$

$$x + 0.69 = 0.71$$

$$x = 0.02$$

$$P(A \cap B \cap C) = 0.02$$

but $\quad P(A) . P(B) . P(C) = 0.5 \times 0.3 \times 0.2 = 0.03$

$$P(A) . P(B) . P(C) \neq P(A \cap B \cap C)$$

$\therefore$ A, B, C are not totally independent.

An alternative method, which requires greater knowledge of the algebra of sets, is as follows:

$$P(A \cup B \cup C) = P(A) + P(B) + P(C) - P(B \cap C) - P(C \cap A) - P(A \cap B) + P(A \cap B \cap C).$$

Since A, B, C are pairwise independent

$$0.71 = 0.5 + 0.3 + 0.2 - 0.3 \times 0.2 - 0.2 \times 0.5 - 0.5 \times 0.3 + P(A \cap B \cap C).$$

$$\therefore \qquad\qquad P(A \cap B \cap C) = 0.02$$

but $\qquad P(A).P(B).P(C) = 0.5 \times 0.3 \times 0.2 = 0.03$

$$P(A).P(B).P(C) \neq P(A \cap B \cap C)$$

$\therefore$ A, B, C are not totally independent.

Miscellaneous Exercise   3

1 The two events A and B are such that

$$P(A) = 0.6, \qquad P(B) = 0.2, \qquad P(A \mid B) = 0.1.$$

Calculate the probabilities that
(i) both events occur
(ii) at least one of the events occurs
(iii) exactly one of the events occurs
(iv) B occurs given that A has occurred.   (JMB)

2 Two events A and B are such that

$$P(A) = 0.4 \text{ and } P(A \cup B) = 0.7.$$

(i) Find the value of $P(A' \cap B)$.
(ii) Find the value of $P(B)$ if A and B are mutually exclusive.
(iii) Find the value of $P(B)$ if A and B are independent.   (JMB)

3 Given $P(A) = 0.3$, $P(B) = 0.4$, $P(A \mid B) = 0.32$, find the probabilities $P(A \cap B)$, $P(A \cup B)$, $P(A \mid B')$, $P(B \mid A)$, $P((A \cup B)')$, $P((A \cap B)')$.

4 Two events A and B are such that

$$P(A) = 0.2, \qquad P(A' \cap B) = 0.22, \qquad P(A \cap B) = 0.18$$

Evaluate   (i) $P(A \cap B')$ (ii) $P(A \mid B)$.

Another event C is such that A and C are mutually exclusive, B and C are independent and $P(C) = 0.3$. Evaluate (iii) $P(A \cup B \cup C)$.   (JMB)

5 Two events A and B are such that

$$P(A) = \frac{1}{2}, \qquad P(B) = \frac{1}{3}, \qquad P(A \mid B) = \frac{1}{4}.$$

Evaluate   (i) $P(A \cap B)$ (ii) $P(A \cup B)$ (iii) $P(A' \cap B')$.

Another event C is such that A and C are independent, and $P(A \cap C) = \frac{1}{12}$,

$P(B \cup C) = \frac{1}{2}$. Show that B and C are mutually exclusive.   (JMB)

6 When a student of English literature is asked to identify the source of a quotation, he has a probability of 0.7 of being able to name the title of the book from which the quotation came, a probability of 0.6 of being able to name the author, and a probability of 0.5 of being able to name both the title and the author of the book.

Calculate the probabilities that the student
(i) will be wrong in both title and author

(ii) will name the correct author but will be wrong in the title.

Given that the student names the correct author, calculate the probability that he will be wrong in the title. (*JMB*)

7 Suppose that for married couples eligible to vote in an election, the probability that a husband will vote is $\frac{3}{5}$, the probability that a wife will vote is $\frac{1}{2}$, and the probability that a wife will vote given that her husband votes is $\frac{2}{3}$.

Calculate the probabilities that, for a randomly chosen married couple,
(i) both will vote

(ii) neither will vote

(iii) only one of them will vote.

If two married couples are chosen at random and they act independently, calculate the probability that exactly one man and exactly one woman will vote. (*JMB*)

8 In a multiple choice question, 6 answers are given, of which only one is correct, and candidates are instructed to select one answer. The probability that a randomly chosen candidate will select the correct answer is p, and all incorrect answers are equally likely.

Find the probabilities that two such candidates
(i) both select the correct answer

(ii) both select the same incorrect answer

(iii) select different answers. (*JMB*)

9 Write down an equation, in terms of probabilities, corresponding to each of the following statements:
(i) the events A and B are independent

(ii) the events A and B are mutually exclusive.

The events A, B and C are such that A and B are independent and A and C are mutually exclusive. Given that $P(A) = 0.4$, $P(B) = 0.2$, $P(C) = 0.3$, $P(B \cap C) = 0.1$ calculate

$$P(A \cup B), \qquad P(C \mid B), \qquad P(B \mid A \cup C).$$

Also calculate the probability that one and only one of the events B, C will occur. (*WJEC*)

10 Events A and B are such that

$$P(A) = P(B) = p, \quad P(A \cap B) = \frac{1}{24} \quad \text{and} \quad P(A' \cap B') = \frac{13}{24}.$$

Show that $p = \frac{1}{4}$ and find $P(A \mid B)$.

A third event C is such that $P(B \cup C) = \frac{7}{12}$ and $P(A \cap B' \cap C') = \frac{1}{8}$. Given that B and C are mutually exclusive, show that A and C are independent.

(*JMB*)

11 What is meant by saying that two events, A and B, are mutually exclusive? What is meant by saying that A and B are independent? Discuss whether two events can be both mutually exclusive and independent.

Explain what is wrong with the following statements, giving an amended calculation where you think it is appropriate.
  (i) Since accident statistics show that the probability that a person will be involved in a road accident in a given year is 0.02, the probability that he will be involved in two accidents in that year is 0.0004.
  (ii) Four persons are asked the same question by an interviewer. If each has, independently, probability $\frac{1}{6}$ of answering correctly, the probability that at least one answers correctly is $4 \times \frac{1}{6} = \frac{2}{3}$. (*JMB*)

12 Alison and Betty are two athletes who run in different races. At a particular athletics meeting, the probability that both girls will win their respective races is $\frac{1}{6}$, while the probability that neither will win is $\frac{1}{4}$.

  a Calculate the probability that only one of the two girls will win.

  b Assuming that the events 'Alison wins' and 'Betty wins' are independent and that Alison has a higher probability of winning than Betty,
   (i) show that the probability that Alison will win her race is $\frac{2}{3}$, and calculate the probability that Betty will win her race
   (ii) calculate the conditional probability that Alison won her race given that only one of the two girls won. (*JMB*)

13 A fair die is thrown twice: A denotes the event that the score obtained on the first throw will be 3 or less; B denotes the event that the two scores obtained will differ by 1 or less; C denotes the event that the sum of the two scores obtained will be 10 or more. Calculate the probability of each of these events A, B and C. Of these three events, show that
  (i) only two of them are mutually exclusive,
  (ii) only two of them are independent. (*JMB*)

14 The faces of a cubical die numbered 1, 2, 3 are coloured red; the faces numbered 4 and 5 are coloured white and the face numbered 6 is coloured blue. Find the probability that on a toss of this die the face uppermost is
   (i) coloured red or numbered 5
   (ii) coloured red or numbered with an odd number
   (iii) numbered 2 given that it is coloured red.

15 A pair of unbiased dice are rolled. Find the probability that
   (i) the sum of the scores is less than 7
   (ii) the sum of the scores is a prime number
   (iii) the sum of the scores is at least 10.

16 In a small club consisting of 10 married couples two persons are to be chosen by lot to be president and secretary. Find the probability that the president and the secretary will be
   (i) both men
   (ii) of the opposite sex
   (iii) married to each other.

17 Assume that, in any family, each child born is equally likely to be a boy or a girl. A family with three children is chosen at random. Find the probability that the oldest child is a girl:
   (i) given no further information
   (ii) given that the family contains exactly one girl
   (iii) given that the family contains exactly two girls
   (iv) given that the family contains at least one girl.   (*JMB*)

18 a Express P($A \cup B$) in terms of P(A) and P(B) in each of the cases when A and B are
   (i) mutually exclusive
   (ii) independent.

   b Suppose that a letter sent by first-class mail has probability 0.6 of being delivered the next day, probability 0.3 of being delivered the second day after the letter was posted, and probability 0.1 of being delivered on the third day after the letter was posted. Suppose, further, that for a letter sent by second-class mail, the corresponding probabilities are 0.2, 0.3 and 0.5 respectively.

   Two letters are posted simultaneously, one by first-class mail and the other by second-class mail. Assuming that delivery days of the letters are independent, calculate the probabilities that
   (i) at least one of the two letters will be delivered the following day
   (ii) the letter sent by first-class mail will be delivered a day earlier than the letter sent by second-class mail.

   If three letters are posted simultaneously by second-class mail, calculate the probability that all three letters will be delivered on the same day.
                                                                      (*WJEC*)

19 An analysis of the other subjects taken by A level Mathematics candidates in a certain year showed that 20 per cent of them took Further Mathematics, 50 per cent took Physics and 5 per cent took both Further Mathematics and Physics. A candidate is chosen at random from those who took Mathematics.

(i) Calculate the probability that the chosen candidate took neither Further Mathematics nor Physics.

(ii) Given that the chosen candidate took at least one of Further Mathematics and Physics, calculate the probability that the candidate took Further Mathematics.

(JMB)

20 The three events A, B, and C are such that

$$P(A) = \frac{3}{5}, \ P(C) = \frac{1}{2}, \ P(B \mid C) = \frac{2}{5}, \ P(B \mid C') = \frac{1}{5}, \ P(B \mid A) = \frac{1}{5}$$

and $P(A \cap B \cap C) = \frac{3}{25}$.

(i) Show that A and B are not independent.

(ii) Show that when both A and B occur together, C also occurs.

(iii) Show that $P(A \cap C) > \frac{1}{5}$.

(JMB)

Chapter 4

Permutations and combinations

As questions about probability become more complex, a knowledge of the elementary properties of permutations and combinations is an invaluable aid.

4.1 Permutations

Multiplication principle

If one operation can be performed in m different ways and a second operation can then be performed in n different ways, then the number of different ways in which the operations may be performed in succession is $m \times n$.

Example 1

Find the number of ways in which a pair of partners for a mixed doubles tennis game may be chosen from a group of 6 boys and 5 girls.

The number of ways in which the boy may be chosen is 6 and the number of ways in which the girl may be chosen is 5, thus the number of ways in which a pair of partners may be chosen is $6 \times 5 = 30$.

The multiplication principle may be extended to include more than two operations.

Example 2

Find the number of elements in the sample space of outcomes when three dice are rolled.

Number of elements $= 6 \times 6 \times 6 = 216$

Permutations

Any ordered arrangement of the elements of a set is called a *permutation* of these elements, e.g. ABC, ACB, BAC, BCA, CAB, CBA are all the possible permutations of the three letters A, B, C.

Example 3

Find the number of permutations of 4 different objects in a row.

Any one of the 4 objects may be placed in the first position; when the first position has been filled, any one of the remaining 3 objects may be placed in the second position. Thus, using the multiplication principle, the first two positions may be filled in 4×3 ways.

When the first two positions have been filled, any one of the remaining 2 objects may be placed in the third position. Thus the first three positions may be filled in $4 \times 3 \times 2$ ways.

When the first three positions have been filled, there is only 1 object left to fill the fourth position. Thus the number of permutations of 4 objects in a row is $4 \times 3 \times 2 \times 1$; this number is denoted by 4! and is called *four factorial* or *factorial four*.

More generally, the number of permutations of n different objects in a row is $n!$, where

$$n! = n(n - 1)(n - 2)\ldots3.2.1.$$

Example 4

Find the number of permutations of 3 objects chosen from 7 different objects in a row.

The first position may be filled in 7 ways; the second position may then be filled in 6 ways and the third position in 5 ways. Thus the number of permutations of 3 objects from 7 objects is $7 \times 6 \times 5 = 210$.

This may be written in an alternative form:

$$7 \times 6 \times 5 = \frac{7 \times 6 \times 5 \times 4 \times 3 \times 2 \times 1}{4 \times 3 \times 2 \times 1} = \frac{7!}{4!}$$

More generally, the number of permutations of r different objects from n different objects is denoted by ${}^nP_r$, where

$${}^nP_r = n(n - 1)(n - 2)\ldots(n - r + 1)$$

or $\qquad {}^nP_r = \dfrac{n!}{(n - r)!}.$

For consistency it is necessary to define 0! to be 1.

Example 5

Find the number of permutations of the 8 letters, A, A, A, B, C, D, E, F.

Consider the 8 letters $A_1, A_2, A_3, B, C, D, E, F$; the number of permutations of these letters is 8!

If any one arrangement of these letters is written down, the letters A_1, A_2, A_3 can be arranged among themselves, without altering the other letters, in 3! ways. If the suffixes of the A's are removed, these 3! arrangements are all identical. Thus the number of arrangements of the 8 letters is $\frac{8!}{3!}$ or 6720.

More generally, the number of permutations of n objects in a row, when there are p alike of one kind, q alike of another kind and r alike of a third kind, is

$$\frac{n!}{p!\,q!\,r!}.$$

Example 6

Find the number of permutations of the letters of the word *MATHEMATICS*.

There are 11 letters in the word *MATHEMATICS*, of which 2 are M's, 2 are A's and 2 are T's. Thus the number of permutations is given by

$$\frac{11!}{2!\,2!\,2!} = 4989600.$$

Exercise 4.1

1 Find the number of permutations of 5 different objects in a row.

2 Find the number of permutations of 43 different objects in a row.

3 Find the number of permutations of 5 different objects, chosen from 8 different objects, in a row.

4 Find the number of permutations of 32 different objects, chosen from 51 different objects, in a row.

5 Find the number of permutations of 4 different objects, chosen from 153 different objects, in a row.

6 In how many ways can a boy and a girl be chosen from 5 boys and 8 girls?

7 In how many ways can a first prize and a second prize be awarded in a class of 30 children? (No child can win both prizes.)

8 In how many ways can two prizes be awarded in a class of 30 children? (Any child may win both prizes.)

9 In how many ways can a batting order be drawn up for a cricket eleven?

10 How many three-digit numbers can be formed using the figures 4, 3, 2, 1
   a if repetitions are allowed
   b if repetitions are not allowed?

11 How many three-digit numbers can be formed using the figures 3, 2, 1, 0
   a if repetitions are allowed
   b if repetitions are not allowed?

12 Find the number of permutations of the letters of the word *MEAN*.

13 Find the number of permutations of the letters of the word *STATISTICS*.

14 Find the number of permutations of the letters of the word *PROBABILITY*.

15 In how many ways can 4 boys and 4 girls be arranged in a row
   a so that the boys and girls are placed alternately
   b with no restriction?

4.2 Combinations

A *combination* of r objects from n different objects is any selection of r of the objects in which the order of selection is irrelevant. The number of combinations of r objects from n different objects is denoted by the symbols

$$^{n}C_{r} \quad \text{or} \quad \binom{n}{r}.$$

Each combination of r objects may be arranged r! ways, thus each combination corresponds to r! permutations:

$$^{n}C_{r} \times r! = {}^{n}P_{r}$$

$$^{n}C_{r} \times r! = \frac{n!}{(n-r)!}$$

$$^{n}C_{r} \quad \text{or} \quad \binom{n}{r} = \frac{n!}{r!(n-r)!}.$$

It is left as an exercise to prove the useful result below.

$$\binom{n}{r} = \binom{n}{n-r}$$

The calculation of $^{n}C_{r}$ for specific values of n and r is often best achieved using the formula

$$\binom{n}{r} = \frac{n(n-1)(n-2)\ldots(n-r+1)}{r(r-1)(r-2)\ldots3.2.1}.$$

Example 7

Find the number of ways in which a squad of 22 players can be selected from a group of 25 players.

$$\text{Number of ways} = \binom{25}{22} = \binom{25}{3}$$

$$= \frac{25 \times 24 \times 23}{3 \times 2 \times 1} = 2300$$

Example 8

Find the greatest number of points of intersection of 7 circles drawn on a plane.

The number of pairs that may be chosen from the 7 circles is $\binom{7}{2}$ which is equal to 21. For each pair of circles there is a maximum of two points of intersection, thus the greatest number of points of intersection of 7 circles is $21 \times 2 = 42$.

Exercise 4.2

1 Calculate the number of combinations of 15 different objects taken 3 at a time.

2 Calculate the number of combinations of 43 different objects taken 17 at a time.

3 Calculate the number of combinations of 87 different objects taken 83 at a time.

4 In how many ways can

a 3 books   **b** 7 books

be selected from 10 different books?

5 In how many ways can a team of 15 boys be chosen from a group of 20 boys?

6 Find the greatest number of points of intersection of 20 straight lines drawn on a plane.

7 Find the greatest number of points of intersection of 10 circles drawn on a plane.

8 Find the greatest number of points of intersection of 20 straight lines and 10 circles drawn on a plane.

9 Find the greatest number of points of intersection of 8 ellipses drawn on a plane.

10 In how many ways can a committee of 3 men and 3 women be chosen from a group of 8 men and 6 women?

11 In how many ways can a group of 10 men be divided into two groups containing 3 and 7 men?

12 In how many ways can a group of 10 men be divided into two groups each containing 5 men?

13 Find the number of
a ordered samples of size 3   **b** unordered samples of size 3
which may be selected, *without* replacement, from the letters a, b, c, d, e.

14 Find the number of
a ordered samples of size 3   **b** unordered samples of size 3
which may be selected, *with* replacement, from the letters a, b, c, d, e.

15 Show that $\binom{n}{r} + \binom{n}{r-1} = \binom{n+1}{r}$.

4.3 Other useful results from Pure Mathematics

In the following chapters some of the exercises may require some results from Pure Mathematics which are unfamiliar to the reader. These exercises may be postponed until later or the following results assumed.

1 If the ratio of successive terms of a series is constant, the series is called a *geometric series* or a *geometric progression* (G.P. is a common abbreviation). Its first term is denoted by a and its common ratio by r. The sum S_n of the first n terms is given by

$$S_n = \frac{a(1 - r^n)}{(1 - r)}$$

and provided that $|r| < 1$, the sum S to infinity is given by

$$S = \frac{a}{(1 - r)}.$$

2 The binomial expansion of $(a + b)^n$, where n is a positive integer, is given by

$$(a + b)^n = \sum_{r=0}^{n} \binom{n}{r} a^{n-r} b^r.$$

3 The binomial expansion of $(1 + x)^n$, where n is not a positive integer and $|x| < 1$, is given by,

$$(1 + x)^n = 1 + nx + \frac{n(n - 1)}{2!} x^2 + \ldots$$

In particular $(1 - x)^{-1} = 1 + x + x^2 + \ldots = \sum_{r=0}^{\infty} x^r$

and $(1 - x)^{-2} = 1 + 2x + 3x^2 + \ldots = \sum_{r=1}^{\infty} r x^{r-1}.$

4 $\displaystyle\sum_{r=1}^{n} r = \frac{n}{2}(n + 1)$

$\displaystyle\sum_{r=1}^{n} r^2 = \frac{n}{6}(n + 1)(2n + 1)$

Miscellaneous Exercise ☐ 4 ☐

1 A squad of 22 football players contains 3 goalkeepers, 8 defenders, 7 midfield players and 4 forwards. Find how many different teams may be selected from the squad if each team includes 1 goalkeeper, 4 defenders, 4 midfield players and 2 forwards.

2 In how many ways can 8 different books be arranged on a shelf if 2 particular books are separated?

3 In how many ways can 2 games of tennis, 4 players in each game, be arranged if 10 players are available?

4 How many even numbers greater than 500 can be formed from the figures 2, 3, 5, 7, 8 if no repetition is allowed?

5 How many different forecasts (win, lose or draw) can be made for the results of 12 matches on a football pool coupon?

6 A pack of 20 cards consists of 5 red, 5 blue, 5 green and 5 yellow cards, and the cards in each set of a particular colour are numbered 1, 2, 3, 4, and 5 respectively. In a certain game, a hand consists of 5 cards. Find numerical expressions for
   (i) the number of different hands which contain cards of at least two different colours,
   (ii) the number of different hands which contain three cards marked with one number and two cards marked with a second number. *(JMB)*

7 Whole numbers are formed from the digits selected from 1, 2, 3, 4, 5, 6 without repetition. Calculate how many of these numbers are less than 2350. *(JMB)*

8 Calculate the greatest number of points of intersection when three circles and four ellipses are drawn on a sufficiently large sheet of paper. *(JMB)*

9 A small chess club consists of 3 married couples, 5 unmarried men and 2 unmarried women. Calculate the number of ways in which a team of four may be chosen in each of the following cases:
(i) when the team is to contain exactly one married couple,

(ii) when the team is to contain at least one man and at least one woman.

[In each case, teams with the same members but chosen in a different order are to be regarded as identical.] *(JMB)*

10 A committee of 6 is to be chosen from 10 men and 5 women; the committee must contain at least 3 men and at least 2 women. Calculate the number of ways in which this committee may be chosen.

If, in addition, two particular women are unable to serve together on the same committee, calculate the number of ways in which the committee may be chosen. *(JMB)*

Chapter 5

More probability

The elementary properties of permutations and combinations together with the rules of probability may be used to answer more difficult questions involving probability. The following worked examples will illustrate a range of techniques useful in problem solving. It is not suggested that the particular method used in each question is the 'best' method, indeed, the 'best' method is often a matter of opinion and personal preference. The reader may wish to apply other methods for comparison purposes.

5.1 Worked examples

Example 1

A committee consists of a chairman, who is a man, five other men and four women.

a A sub-committee of four is to be chosen at random from the committee members. Find the probabilities that the sub-committee will consist of (i) only men, (ii) two men and two women.

b Given that the chairman must serve on the sub-committee and that the other three members are to be chosen as before, find the probability that the sub-committee will consist of two men and two women.

a Number of possible sub-committees $= \binom{10}{4} = 210$.

(i) Number of ways of choosing four men for the sub-committee $= \binom{6}{4} = 15$.

Probability required $= \dfrac{15}{210} = \dfrac{1}{14}$.

(ii) Number of ways of choosing two men for the sub-committee $= \binom{6}{2} = 15$.

Number of ways of choosing two women for the sub-committee $= \binom{4}{2} = 6$.

Number of ways of choosing two men and two women $= 15 \times 6 = 90$.

Probability required $= \dfrac{90}{210} = \dfrac{3}{7}$.

71

b Number of possible sub-committees $= \binom{9}{3} = 84.$

Number of ways of choosing one extra man and two women $= \binom{5}{1} \times \binom{4}{2} = 30.$

Probability required $= \dfrac{30}{84} = \dfrac{5}{14}.$

Example 2

A bag contains 6 red balls, 4 blue balls and 2 white balls.

a If three balls are drawn at random from the bag, find the probability that two are red and one is white.

b If five balls are drawn at random from the bag, find the probability that at least three are red.

---

a Let R_1 denote the event that the first ball drawn is red, R_2 the event that the second ball drawn is red, W_3 the event that the third ball drawn is white.

$$P(R_1 \cap R_2 \cap W_3) = P(R_1) . P(R_2 \cap W_3 \mid R_1)$$

$$P(R_1 \cap R_2 \cap W_3) = P(R_1) . P(R_2 \mid R_1) . P(W_3 \mid R_1 \cap R_2)$$

Since there are 6 red balls among the 12 balls in the bag $\qquad P(R_1) = \dfrac{6}{12}.$

Since there are 5 red balls among the remaining 11 balls $\qquad P(R_2 \mid R_1) = \dfrac{5}{11}.$

Since there are 2 white balls among the remaining 10 balls $P(W_3 \mid R_1 \cap R_2) = \dfrac{2}{10}.$

$$\therefore \quad P(R_1 \cap R_2 \cap W_3) = \frac{6}{12} \times \frac{5}{11} \times \frac{2}{10} = \frac{1}{22}$$

Similarly $\qquad P(R_1 \cap W_2 \cap R_3) = \dfrac{6}{12} \times \dfrac{2}{11} \times \dfrac{5}{10} = \dfrac{1}{22}$

and $\qquad P(W_1 \cap R_2 \cap R_3) = \dfrac{2}{12} \times \dfrac{6}{11} \times \dfrac{5}{10} = \dfrac{1}{22}.$

(N.B. Although the individual fractions in the products above are different, each product is equal to $\dfrac{1}{22}$.)

Since the three compound events above are mutually exclusive

P(2 red and 1 white balls drawn) $= \dfrac{1}{22} + \dfrac{1}{22} + \dfrac{1}{22} = \dfrac{3}{22}.$

Although the notation used above is clear and, to a large extent, self-explanatory, it is rather cumbersome. An alternative layout of the same method is shown below.

The symbols RRW in order are used to denote the event $R_1 \cap R_2 \cap W_3$

$$P(RRW) = \frac{6}{12} \times \frac{5}{11} \times \frac{2}{10} = \frac{1}{22}.$$

The number of ways of arranging the symbols RRW in a row $= \frac{3!}{2!} = 3$

$$P(2 \text{ red and } 1 \text{ white balls drawn}) = \frac{3}{22}.$$

This may be further reduced to

P(2 red and 1 white balls drawn) = P(RRW in any order)

$$= \frac{6}{12} \times \frac{5}{11} \times \frac{2}{10} \times \frac{3!}{2!} = \frac{3}{22}.$$

b This part of the question is concerned with the redness or otherwise of the balls drawn.

P(exactly 3 red balls drawn) = P($RRRR'R'$ in any order)

$$= \frac{6}{12} \times \frac{5}{11} \times \frac{4}{10} \times \frac{6}{9} \times \frac{5}{8} \times \frac{5!}{3!\,2!}$$

$$= \frac{50}{132}$$

P(exactly 4 red balls drawn) = P($RRRRR'$ in any order)

$$= \frac{6}{12} \times \frac{5}{11} \times \frac{4}{10} \times \frac{3}{9} \times \frac{6}{8} \times \frac{5!}{4!}$$

$$= \frac{15}{132}$$

P(exactly 5 red balls drawn) = P($RRRRR$ in any order)

$$= \frac{6}{12} \times \frac{5}{11} \times \frac{4}{10} \times \frac{3}{9} \times \frac{2}{8} \times \frac{5!}{5!}$$

$$= \frac{1}{132}$$

$$P(\text{at least 3 red balls drawn}) = \frac{50}{132} + \frac{15}{132} + \frac{1}{132}$$

$$= \frac{66}{132} = \frac{1}{2}$$

Example 3

In one form of poker a hand of 5 cards is dealt to each player from a well shuffled pack of 52 playing-cards. Calculate the probabilities that a player will be dealt

a   a full house (3 cards of one kind and 2 cards of another, e.g. 3 kings and 2 aces)

b   a flush (5 cards of the same suit)

c   two pairs.

a   P(a particular full house) = P($KKKAA$ in any order)

$$= \frac{4}{52} \times \frac{3}{51} \times \frac{2}{50} \times \frac{4}{49} \times \frac{3}{48} \times \frac{5!}{3!\,2!}$$

$$= \frac{1}{108290}$$

Number of ways of choosing a kind of card for the trio   = 13
Number of ways of then choosing a kind of card for the pair = 12
Number of types of full house = 13 × 12 = 156

P(a full house) $= 156 \times \dfrac{1}{108290}$

$$= \frac{6}{4165}$$

b   P(a particular flush)   = P($HHHHH$ in any order)

$$= \frac{13}{52} \times \frac{12}{51} \times \frac{11}{50} \times \frac{10}{49} \times \frac{9}{48}$$

$$= \frac{33}{66640}$$

Number of types of flush = 4

P(a flush) $= 4 \times \dfrac{33}{66640}$

$$= \frac{33}{16660}$$

c   P(a particular 2 pair hand) = P($AAKKQ$ in any order)

$$= \frac{4}{52} \times \frac{3}{51} \times \frac{4}{50} \times \frac{3}{49} \times \frac{4}{48} \times \frac{5!}{2!\,2!}$$

$$= \frac{3}{54145}$$

Number of ways of choosing the kinds of card for the 2 pairs $= \binom{13}{2} = 78$

Number of ways of then choosing the kind of card for single card $= 11$

Number of types of 2 pair hands $= 78 \times 11 = 858$

$$P(2 \text{ pair hand}) = 858 \times \frac{3}{54145}$$

$$= \frac{198}{4165}$$

Example 4

A fair coin is tossed 10 times. Find the probability that there will be at least 5 heads with exactly 5 of them occurring consecutively.

Consider the case when the 5 heads occur first and are immediately succeeded by a tail

$$HHHHHT\,X\,X\,X\,X$$

the remaining 4 tosses, denoted by X, may occur in $2 \times 2 \times 2 \times 2 = 16$ ways.

Similarly, the case when the 5 heads occur last and are immediately preceded by a tail may occur in 16 ways.

Now consider the cases when the 5 heads are immediately preceded by a tail and are immediately succeeded by a tail, e.g.

$$THHHHHT\,X\,X\,X$$

the remaining 3 tosses may occur in $2 \times 2 \times 2 = 8$ ways; thus there are 8 ways in which this particular arrangement may occur.

Similarly there are 8 ways in which each of the following arrangements may occur

$$X\,THHHHHT\,X\,X, \quad X\,X\,THHHHHT\,X, \quad X\,X\,X\,THHHHHT.$$

The total number of ways in which 'at least 5 heads with exactly 5 of them consecutive' may occur is $16 + 16 + 8 + 8 + 8 + 8 = 64$.

The total number of possible outcomes of 10 tosses $= 2^{10} = 1024$

$$P(\text{required event}) = \frac{64}{1024} = \frac{1}{16}$$

Example 5

In a certain game, a person throws a pair of dice and he wins on his first throw if he throws a total score of 7 or 11; he loses on his first throw if he throws a total score of 2, 3 or 12; if he throws a total score of 4, 5, 6, 8, 9 or 10 he continues to throw the dice either until he throws the same total score as he threw on his first throw, in which case he wins, or until he throws a total score of 7, in which case he loses. Find the probability that he wins by throwing a total score of 4 on his first throw and a total score of 4 again subsequently.

Let F denote the event 'throwing a total score of 4 on one throw of the dice'.

$$P(F) = P((1, 3) \text{ or } (2, 2) \text{ or } (3, 1)) = \frac{3}{36} = \frac{1}{12}$$

Let X denote the event 'not throwing a total score of 4 or 7 on one throw'.

$$P(\text{score} = 7) = P((1, 6) \text{ or } (2, 5) \text{ or } (3, 4) \text{ or } (4, 3) \text{ or } (5, 2) \text{ or } (6, 1)) = \frac{6}{36}$$

$$P(X) = 1 - \frac{3}{36} - \frac{6}{36} = \frac{3}{4}$$

The player will win in the manner described only if one of the following successions of events occurs:

$$FF \text{ or } FXF \text{ or } FXXF \text{ or } FXXXF \text{ or } \dots.$$

$$P(FF) = \frac{1}{12} \times \frac{1}{12}$$

$$P(FXF) = \frac{1}{12} \times \frac{3}{4} \times \frac{1}{12}$$

$$P(FXXF) = \frac{1}{12} \times \frac{3}{4} \times \frac{3}{4} \times \frac{1}{12} \quad \text{and so on.}$$

Since the events $FF, FXF, FXXF, \dots$ are mutually exclusive,

$$P(\text{required event}) = P(FF) + P(FXF) + P(FXXF) + \dots$$

$$= \frac{1}{144} + \frac{1}{144} \times \frac{3}{4} + \frac{1}{144} \times \left(\frac{3}{4}\right)^2 + \dots.$$

This is an infinite G.P. with first term $a = \dfrac{1}{144}$ and common ratio $r = \dfrac{3}{4}$. Using the formula $S = \dfrac{a}{(1 - r)}$. [See Result 1 on page 68.]

$$P(\text{required event}) = \frac{\dfrac{1}{144}}{\left(1 - \dfrac{3}{4}\right)} = \frac{1}{36}$$

Example 6

In a biological experiment, a large batch of seeds is produced. It is known that the batch contains three types of seed A, B and C in the ratios $5:3:2$. Eighty per cent of type A seeds are fertile, sixty per cent of type B seeds are fertile and forty per cent of type C seeds are fertile. A seed is chosen at random from the batch. Calculate

a the probability that the seed is fertile

b the probability that the seed is of type A, given that it is fertile.

Whenever a random experiment may be considered to consist of a sequence of stages a probability tree diagram may facilitate the calculation of associated probabilities. In a tree diagram, the branches indicate all the possibilities at each stage and probabilities may be attached to the various branches.

a

Type	Fertile/Infertile	Outcome	Probability

0.8	F	$A \cap F$	$0.5 \times 0.8 = 0.40$
0.2	F'	$A \cap F'$	$0.5 \times 0.2 = 0.10$
0.6	F	$B \cap F$	$0.3 \times 0.6 = 0.18$
0.4	F'	$B \cap F'$	$0.3 \times 0.4 = 0.12$
0.4	F	$C \cap F$	$0.2 \times 0.4 = 0.08$
0.6	F'	$C \cap F'$	$0.2 \times 0.6 = 0.12$

$$P(F) = P(A \cap F) + P(B \cap F) + P(C \cap F)$$

$$P(F) = 0.40 + 0.18 + 0.08$$

$$P(F) = 0.66$$

b $P(A \mid F) = \dfrac{P(A \cap F)}{P(F)}$

$$= \frac{0.40}{0.66} = \frac{20}{33}$$

The answer to **Example 6a** above is an example of the application of the total probability rule, which is stated below.

5.2 Total probability rule

If $(A_1, A_2, \ldots, A_n)$ is a set of mutually exclusive and exhaustive events associated with a random experiment and B is another event associated with the experiment, then:

$$P(B) = \sum_{i=1}^{n} P(A_i) . P(B \mid A_i).$$

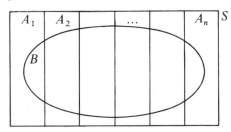

Since the events $A_1, A_2, \ldots, A_n$ are exhaustive:

$$A_1 \cup A_2 \cup \ldots \cup A_n = S.$$

Hence $B = B \cap S = B \cap (A_1 \cup A_2 \cup \ldots \cup A_n)$

and $B = (B \cap A_1) \cup (B \cap A_2) \cup \ldots \cup (B \cap A_n).$

The events $(B \cap A_1), (B \cap A_2), \ldots, (B \cap A_n)$ are mutually exclusive, therefore

$$P(B) = \sum_{i=1}^{n} P(B \cap A_i) \qquad \text{(Rule 3)}$$

$$P(B) = \sum_{i=1}^{n} P(A_i) . P(B \mid A_i). \qquad \text{(Rule 5)}$$

The result in **Example 6b** might have been obtained using Bayes' theorem, which is stated below.

5.3 Bayes' theorem

If $\{A_1, A_2, \ldots, A_n\}$ is a set of mutually exclusive and exhaustive events associated with a random experiment and B is another event associated with the experiment, then:

$$P(A_j \mid B) = \frac{P(A_j) . P(B \mid A_j)}{\sum\limits_{i=1}^{n} P(A_i) . P(B \mid A_i)} \qquad \text{for } j = 1, 2, \ldots, n.$$

Using Rule 5: $P(A_j \mid B) = \dfrac{P(A_j \cap B)}{P(B)}$

Using Rule 5: $P(A_j \mid B) = \dfrac{P(A_j) . P(B \mid A_j)}{P(B)}.$

Using the total probability rule:
$$P(A_j | B) = \frac{P(A_j) \cdot P(B | A_j)}{\sum\limits_{i=1}^{n} P(A_i) \cdot P(B | A_i)}.$$

In many easy examples there is no need to refer to Bayes' theorem explicitly.

Example 7

Three factories A, B, C produce large numbers of identical electronic components in the ratios 7:2:1. The proportions of defective components produced by A, B, C are 0.003, 0.004 and 0.005 respectively. Given that a component selected at random from the pooled output of the factories is defective, calculate the probability that it was produced in factory C.

Using the total probability rule:
$$P(D) = P(A) \cdot P(D | A) + P(B) \cdot P(D | B) + P(C) \cdot P(D | C)$$
$$P(D) = 0.7 \times 0.003 + 0.2 \times 0.004 + 0.1 \times 0.005$$
$$P(D) = 0.0021 + 0.0008 + 0.0005$$
$$P(D) = 0.0034.$$

Using Bayes' theorem or Rule 5:
$$P(C | D) = \frac{P(C) \cdot P(D | C)}{P(D)}$$
$$P(C | D) = \frac{0.0005}{0.0034} = \frac{5}{34}.$$

Example 8

Two dice are to be thrown n times. Find the minimum value of n if the probability of throwing at least one double in the n throws is to exceed 0.9.

P(double on one throw) $= \dfrac{6}{36} = \dfrac{1}{6}$

P(no double on one throw) $= 1 - \dfrac{1}{6} = \dfrac{5}{6}$

P(no double in n throws) $= \left(\dfrac{5}{6}\right)^n$

P(at least one double in n throws) $= 1 - \left(\dfrac{5}{6}\right)^n$

Therefore $1 - \left(\dfrac{5}{6}\right)^n > 0.9$

$$0.1 > \left(\dfrac{5}{6}\right)^n.$$

By trial $n = 13.$

An alternative method, using logarithms, may be used for the last part.

$$0.1 > \left(\frac{5}{6}\right)^n$$

$$\therefore \quad \log(0.1) > \log\left(\frac{5}{6}\right)^n$$

$$\therefore \quad \log(0.1) > n\log\left(\frac{5}{6}\right)$$

Dividing both sides of the inequation by the negative number $\log\left(\frac{5}{6}\right)$

gives

$$\frac{\log(0.1)}{\log\left(\frac{5}{6}\right)} < n$$

$$\therefore \quad n > 12.6$$

$$\therefore \quad \text{minimum value of } n = 13.$$

Various problems in probability arise from geometric considerations. The following example is a simple illustration.

Example 9

An archer shoots arrows at a circular target of radius 75 cm which is marked with two concentric circles of radii 25 cm and 50 cm. When the arrow hits the target at a point which lies
  (i) inside the smaller circle, the archer scores 10 points,
 (ii) between the two circles, he scores 5 points,
(iii) between the larger circle and the edge of the target, he scores 1 point.

Given that the archer always hits the target and that each point of the target is equally likely to be hit on each occasion, find the probability that the archer scores 16 points when he shoots three arrows.

Since each point of the target is equally likely to be hit, the probability of hitting a particular region is proportional to the area of that region.

Let X be the score when he shoots one arrow.

$$P(X = 10) = \frac{\text{area of the smaller circle}}{\text{area of the target}} = \frac{\pi \times 25^2}{\pi \times 75^2} = \frac{1}{9}$$

$$P(X = 5) = \frac{\text{area between two circles}}{\text{area of the target}} = \frac{\pi \times 50^2 - \pi \times 25^2}{\pi \times 75^2} = \frac{1}{3}$$

$$P(X = 1) = \frac{\text{area outside the larger circle}}{\text{area of the target}} = \frac{\pi \times 75^2 - \pi \times 50^2}{\pi \times 75^2} = \frac{5}{9}$$

Let S be the total score when he shoots three arrows.

$$P(S = 16) = P(10, 5, 1 \text{ in any order})$$

$$= \frac{1}{9} \times \frac{1}{3} \times \frac{5}{9} \times 3!$$

$$= \frac{10}{81}$$

In the exercises that follow it is essential to remember the rules of probability which are listed below.

Rule 1 $P(A') = 1 - P(A)$

Rule 2 $P(A \cup B) = P(A) + P(B) - P(A \cap B)$

Rule 3   If A and B are mutually exclusive, then $P(A \cup B) = P(A) + P(B)$.

Rule 4 $P(A \cap B') = P(A) - P(A \cap B)$

Rule 5 $P(B \mid A) = \dfrac{P(A \cap B)}{P(A)}$ or $P(A \cap B) = P(A) \cdot P(B \mid A)$.

Rule 6 $P(A \cap B) = P(A) \cdot P(B)$ if and only if A and B are independent.

Rule 7   If $\{A_1, A_2, \ldots, A_n\}$ is a set of mutually exclusive and exhaustive events, then

$$P(B) = P(A_1) \cdot P(B \mid A_1) + P(A_2) \cdot P(B \mid A_2) + \ldots + P(A_n) \cdot P(B \mid A_n).$$

Miscellaneous Exercise $\boxed{5}$

1 Three cards are drawn at random without replacement from a pack of ten cards which are numbered from 1 to 10 respectively. Calculate
   (i) the probability that the numbers drawn consist of two even numbers and one odd number,
   (ii) the probability that at least one of the numbers drawn is a perfect square greater than 1,
   (iii) the probability that the smallest number drawn is the 5.   (*JMB*)

2 An unbiased cube has one face coloured red, two faces coloured blue, and its remaining three faces coloured green. If the cube is thrown four times, calculate the probabilities of
   (i) blue appearing uppermost at least once,
   (ii) blue appearing uppermost at least twice,
   (iii) red appearing uppermost exactly twice,
   (iv) both (ii) and (iii),
   (v) at least one of (ii) and (iii).   (*JMB*)

3 Four cards are drawn at random, without replacement, from a pack of nine cards which are numbered from 1 to 9 respectively.
  a Calculate the probability that
    (i) both the numbers 1 and 9 will be drawn,
    (ii) the largest number drawn will be 8,
    (iii) at least three even numbers will be drawn.

  b Given that the largest number drawn was 8, calculate the conditional probability that the smallest number drawn was 3.

  c If the drawn cards are set down in a row from left to right as they are drawn, calculate the probability that the resulting four digit number will be less than 5941.
  (*WJEC*)

4 Three balls are drawn at random without replacement from a box containing 8 white, 4 black and 4 red balls. Calculate the probabilities that they will consist of
  (i) at least one white ball,
  (ii) two white balls and one black ball,
  (iii) two balls of one colour and the other of a different colour,
  (iv) one ball of each colour.

  If, instead, each ball is replaced in the box before the next ball is drawn, calculate the probability that the three balls drawn will consist of one of each colour.
  (*WJEC*)

5 Three fair dice are thrown. Determine the probabilities that
  (i) none of the individual scores is greater than 4,
  (ii) the individual scores on the three dice are consecutive numbers (irrespective of order),
  (iii) the total score on the three dice is greater than 4, and
  (iv) the highest individual score is 4.
  (*JMB*)

6 An unbiased die is thrown six times. Calculate the probabilities that the six scores obtained will
  (i) consist of exactly two 6's and four odd numbers,
  (ii) be 1, 2, 3, 4, 5, 6 in some order,
  (iii) have a product which is an even number,
  (iv) be such that a 6 occurs only on the last throw and that exactly three of the first five throws result in odd numbers.
  (*JMB*)

7 In a game of chance each player throws two unbiased dice and scores the difference between the larger and smaller numbers which arise. Two players compete and one or the other wins if, and only if, he scores at least 4 more than his opponent. Find the probability that neither player wins.
  (*JMB*)

8 In a biological experiment large numbers of three types of seed A, B and C are produced in the ratios $5:3:2$. Type A seeds are always fertile, type C seeds are always sterile, whilst on average one third of type B are fertile, the remaining two thirds being sterile.
  (i) Find the probability that a seed chosen at random from those produced in the experiment is sterile.
  (ii) Four such seeds are chosen at random. Find the probability that at least three are fertile.
  (iii) Determine how large a random sample of seeds needs to be chosen for there to be a probability of at least 0.99 that the sample contains at least one fertile seed.

  (JMB)

9 Three girls, two of whom are sisters, and five boys, two of whom are brothers, meet to play tennis. They draw lots to determine how they should split up into two groups to play doubles.

  a Calculate the probabilities that one of the two groups will consist of
  (i) boys only,
  (ii) two boys and two girls,
  (iii) the two brothers and the two sisters.

  b If the lottery is organised so as to ensure that one of the two groups consists of two boys and two girls, calculate the probability that the two brothers and the two sisters will be in the same group. Given that the two brothers are in the same group, calculate the probability that the two sisters are also in that group.

  (WJEC)

10 a Calculate (to two decimal places) the probabilities of obtaining
  (i) at least one six in 4 throws of an unbiased die,
  (ii) at least one double six in 24 throws of two unbiased dice.

  b Ten people are in a room. If their birthdays are, independently, equally likely to fall in any month, determine the probabilities (leaving your results in factorised form, if you wish)
  (i) that they all fall in different months,
  (ii) that they all fall in March or April, with at least one in each month.

  (JMB)

11 A fair coin is tossed $n(>1)$ times. Let A denote the event that there will be at least one head and at least one tail, and let B denote the event that there will be at most one head. Show that A and B are independent if $n = 3$, but are not independent for any other $n > 1$.

Determine the minimum value of n if the probability of throwing at least one head and at least one tail in n tosses is to exceed 0.9.

  (WJEC)

12 Two rooms L and R each contain p calculating machines; one of the machines in L is damaged; all the others are undamaged. During one morning, q (less than p) of the machines in L are chosen at random and moved to R. During the afternoon of the same day q of the machines in R are chosen at random and moved to L. Find, in the simplest form, the probability that the damaged machine is in L

  (i) at the end of the morning,

  (ii) at the end of the afternoon.   (*JMB*)

13 In a class of 30 pupils, 12 walk to school, 10 travel by bus, 6 cycle and 2 travel by car. If 4 pupils are chosen at random, obtain the probabilities that

  (i) they all travel by bus,

  (ii) they all travel by the same means.

If 2 pupils are picked at random from the class, find the probability that they travel by different means.

In picking out the pupils from the class, find the probability that more than three trials are necessary before a pupil who walks to school is selected.   (*JMB*)

14 A domino is an object marked with two numbers from 0 to 6; the two numbers may be equal or different. Dominoes on which the two numbers are equal are called 'doubles'. A set contains one example of each distinct domino, including a domino with two zeros.

Show that there are 28 dominoes in a set. State the probabilities that a domino drawn at random from the set

  (i) has at least one six in its pair of numbers,

  (ii) is a double six,

  (iii) is a double.

Two fair six-sided dice are thrown. Find the probability that at least one of the dice has a six uppermost.

A domino is drawn at random and the two dice are thrown. Obtaining your answers as exact fractions, find the probabilities

  (iv) that at least one six occurs on the domino and on at least one of the dice,

  (v) that the numbers on the two dice are the same as those on the domino,

  (vi) that at least one number is common to the domino and at least one of the dice.   (*JMB*)

15 A box A contains 10 balls, 4 of which are white and 6 black. If 3 balls are drawn at random without replacement from the box calculate the probabilities that they will consist of

  (i) three balls having the same colour,

  (ii) more white than black balls.

Suppose that 2 balls are drawn at random from the original 10 balls in A and are placed in box B, which originally contained 6 balls, 3 of which were white and 3 were black. If a ball drawn at random from the 8 balls which are in B is found to be black calculate the probability that the 2 balls that were drawn from A and placed in B were both black. *(WJEC)*

16 a Two events A and B are such that $P(A) = \frac{1}{3}$ and $P(B) = \frac{1}{2}$. If A' denotes the complement of A, calculate $P(A' \cap B)$ in each of the cases when

   (i) $P(A \cap B) = \frac{1}{8}$,
   (ii) A and B are mutually exclusive,
   (iii) A is a subset of B.

   b A Scottish court may give any one of the three verdicts: 'guilty', 'not guilty' and 'not proven'. Of all cases tried by the court, 70 per cent of the verdicts are 'guilty', 20 per cent are 'not guilty', and 10 per cent are 'not proven'. Suppose that when the court's verdict is 'guilty', 'not guilty' and 'not proven', the probabilities that the accused is really innocent are 0.05, 0.95 and 0.25, respectively. Calculate the probability that an innocent person will be found 'guilty' by the court. *(WJEC)*

17 Each of two boxes contains ten discs. In one box four of the discs are red, two are white and four are blue; in the other box two are red, three are white and five are blue. One of these boxes is chosen at random and three discs are drawn at random from it without replacement. Calculate
   (i) the probability that one disc of each colour will be drawn,
   (ii) the probability that no white disc will be drawn,
   (iii) the most probable number of white discs that will be drawn.
Given that three blue discs were drawn, calculate the conditional probability they came from the box that contained four blue discs.

   (WJEC)

18 A pack of eight cards consists of four aces and four kings from a pack of ordinary playing cards.
   a Two cards are dealt at random from this pack of eight cards.
      (i) Given that at least one of the two cards dealt is an ace, calculate the probability that both cards are aces.
      (ii) Given that one of the two cards dealt is the ace of spades, calculate the probability that the other card is an ace.

   b Suppose now that the eight cards are shuffled and are dealt one after the other.
      (i) Calculate the probability that the fifth card dealt will be the fourth ace dealt.
      (ii) Calculate the probability that the four aces will be dealt consecutively. *(WJEC)*

19 Each of two bags A and B contains five white and four black balls, while a third bag C contains three white and six black balls.
  (i) Suppose that one of the three bags was chosen at random and that two balls drawn at random without replacement were both black. Calculate the probability that the chosen bag was C.

  (ii) Suppose, instead, that two of the three bags were chosen at random and that one ball was drawn at random from each of the bags. Given that both balls drawn were black, calculate the probability that C was one of the chosen bags. (*WJEC*)

20 Four persons are chosen at random from a group of ten persons consisting of four men and six women. Three of the women are sisters. Calculate the probabilities that the four persons chosen will
  (i) consist of four women,

  (ii) consist of two women and two men,

  (iii) include the three sisters. (*JMB*)

21 Alec and Bill frequently play each other in a series of games of table tennis. Records of the outcomes of these games indicate that whenever they play a series of games, Alec has probability 0.6 of winning the first game and that in every subsequent game in the series, Alec's probability of winning the game is 0.7 if he won the preceding game but only 0.5 if he lost the preceding game. A table tennis game cannot be drawn. Find the probability that Alec will win the third game in the next series of games played against Bill. (*JMB*)

22 Four cards are to be dealt at random, without replacement, from a pack of ten cards, of which two are red and eight are black. Find the most probable number of red cards that will be dealt. (*JMB*)

23 Four cards are to be drawn at random without replacement from a pack of cards numbered from 1 to 10 respectively.
  a Calculate the probabilities that
    (i) the largest number drawn will be 6,

    (ii) the product of the four numbers drawn will be even,

    (iii) all four numbers drawn will be consecutive integers.

  b Given that at least two of the four numbers drawn were even, find the probability that every number drawn was even. (*WJEC*)

24 a Two independent events A and B are such that $P(A) = 0.4$ and $P(A \cup B) = 0.7$. Evaluate $P(B)$ and $P(A' \cap B)$.

  b In order to estimate what proportion of pupils at a school smoked it was decided to interview a random sample of pupils. To encourage truthful answers each pupil in the sample was asked to toss a coin and, without divulging the outcome to the interviewer, to answer 'yes' or 'no' to one of two questions dependent on the outcome of the toss. If the pupil had tossed a head the question to be answered was 'Is your birthday in April?', while if the pupil had tossed a tail the question to be answered was 'Do you smoke?'. It is known that the proportion of the pupils who were born in the month of April is 0.1. Given that the proportion of the sampled pupils that answered 'yes' was 0.2, estimate the proportion of pupils in the school who smoke. (*WJEC*)

25 Four ballpoint refills are to be drawn at random without replacement from a bag containing ten refills, of which 5 are red, 3 are green and 2 are blue. Find
   (i) the probability that both the blue refills will be drawn,
   (ii) the probability that at least one refill of each colour will be drawn. (*JMB*)

26 a The three events A, B, C have respective probabilities $\frac{2}{5}, \frac{1}{3}$ and $\frac{1}{2}$. Given that A and B are mutually exclusive, $P(A \cap C) = \frac{1}{5}$ and $P(B \cap C) = \frac{1}{4}$,
   (i) show that only two of the three events are independent,
   (ii) evaluate $P(C \mid B)$ and $P(A' \cap C')$.

   b When Alec, Bert and Chris play a particular game their respective probabilities of winning are 0.3, 0.1 and 0.6, independently for each game played. They agree to play a series of up to five games, the winner of the series (if any) to be the first player to win three games. Given that Bert wins the first two games of the series show that
   (i) Bert is just over 10 times more likely than Alec to win the series,
   (ii) there is a slightly better than even chance that there will be a winner to the series. (*WJEC*)

27 The probabilities of A, B and C winning a certain game in which all three take part are 0.5, 0.3 and 0.2 respectively. A match is won by the player who first wins two games. Find the probability that A will win a match involving all three players.

When the players are joined by a fourth player, D, the probabilities of A, B or C winning a game are reduced to 0.3, 0.2 and 0.1, respectively. A match is played with all four players taking part in each game; again, the player who first wins two games wins the match. Find the probabilities that D wins in fewer than
   (i) four games,
   (ii) five games,
   (iii) six games. (*JMB*)

28 A machine contains four components, *A*, *B*, *C* and *D*, each of which, independently of the others, may fail with probabilities 0.1, 0.3, 0.4 and 0.2 respectively, when the machine is switched on. Show that there is a probability of 0.88 that at least one of *B* and *C* will not fail.

The machine will fail to operate if *A* fails or if *D* fails or if both *B* and *C* fail, but otherwise it will operate. Calculate the probability that the machine will operate.

A second machine also contains the four components *A*, *B*, *C* and *D*, with the same failure probabilities as above, except that if either *B* or *C* fails the probability of *D* failing is increased to 0.4. Assuming that this machine will fail to operate under the same conditions as those for the first machine, calculate the probability that the second machine will operate.

(*JMB*)

29 A box contains twelve balls numbered from 1 to 12. The balls numbered 1 to 5 are red, those numbered 6 to 9 are white, and the remaining three balls are blue. Three balls are to be drawn at random without replacement from the box. Let *A* denote the event that each number drawn will be even, *B* the event that no blue ball will be drawn, and *C* the event that one ball of each colour will be drawn. Calculate
  (i) P(*A*),   (ii) P(*B*),   (iii) P(*C*),
  (iv) P(*A* ∩ *C*),   (v) P(*B* ∪ *C*),   (vi) P(*A* ∪ *B*).   (*WJEC*)

30 Three numbers are chosen at random without replacement from the set
(1, 2, 3, ..., 10}. Calculate the probabilities that for the three numbers drawn
  (i) none will be greater than 7,
  (ii) the smallest will be 7,
  (iii) their sum will be 7.   (*JMB*)

31 a In each of the following two games show that the probability that the player concerned will win is $\frac{1}{2}$.
  Game 1:   The player throws two fair dice together and wins if he throws at least one 5 or if the sum of the two scores is equal to 6 or equal to 7.
  Game 2:   The player tosses a fair coin four times in succession and wins if he tosses at least two heads successively.
  b The three events *A*, *B* and *C* are such that *A* and *B* are mutually exclusive, *A* and *C* are independent, and P(*A*) = 0.2, P(*B*) = 0.1, P(*A* ∪ *C*) = 0.5, P(*B* ∪ *C*) = 0.4.
  (i) Evaluate P(*C*).
  (ii) Determine whether *B* and *C* are independent, mutually exclusive or neither.   (*WJEC*)

32 Each of three boxes *A*, *B* and *C* contains four balls. Each of the four balls in *A* are red. Two of the four balls in *B* are red and the remaining two balls are white. Three of the four balls in *C* are white and the remaining ball is yellow. Two fair coins are tossed together. If two heads are tossed, one ball is drawn at random from box A; if two tails are tossed, one ball is drawn at random from box B; otherwise, one ball is drawn at random from box C.

(i) Show that the probability of the drawn ball being white is four times that of it being yellow.

(ii) Given that the drawn ball is red, find the conditional probability that it came from box B.

The box from which the ball was drawn is then set aside, and a ball is drawn at random from one of the other two boxes, the choice of box being determined by the outcome of the toss of a fair coin.

(iii) Calculate the probability that the two balls drawn are of the same colour.

(*WJEC*)

33 At a certain school a Recreation Committee comprises three representatives from each of four school houses, White House, Red House, Yellow House and Blue House. A sub-committee of four pupils is randomly selected from this committee to organise the school sports day. Show that the number of different sub-committees that can be formed is 495.

Find the probabilities that
(i) no one is selected from White House,
(ii) one pupil is selected from each house,
(iii) one pupil is selected from White House, two from Red House and one from either of the other two houses.

Given that there are five girls on the Recreation Committee, find the probability that just one girl is on the sub-committee.

It is later decided that the Head Prefect, who is a White House representative on the Recreation Committee, must be on the sub-committee. In this case find the probability that each house will be represented on the sub-committee when the other three members are randomly chosen.

(*JMB*)

34 a Four cards are to be drawn at random without replacement from a pack of nine cards which are numbered from 1 to 9, respectively. Calculate the probabilities that the numbers on the four cards drawn will be such that
(i) the highest of them is 8,
(ii) two of them are 1 and 9,
(iii) at least three of them are odd numbers,
(iv) their product is exactly divisible by 9.

b A local council has 40 members, of whom 30 are male and 24 are over 60 years old. One member is chosen at random. Let A denote the event that the chosen member is male and let B denote the event that the chosen member is over 60 years old. Given that A and B are independent, determine the number of male members over 60 years old.

(*WJEC*)

35 A box A contains 3 red balls and 2 white balls, while a box B contains 4 red balls and 1 white ball. A fair die is thrown. If the score obtained is 3 or 6, a ball is drawn at random from box A and placed in box B, while if the score is neither 3 nor 6 a ball is drawn at random from box B and placed in box A. Calculate the probability that the ball transferred from one box to the other is a red one.

After the transfer of the ball, a ball is drawn at random from the box that contains the extra ball. Show that the probability that both balls drawn are of the same colour is equal to $\dfrac{19}{30}$.

Given that both balls drawn were of the same colour, calculate the conditional probabilities that
(i) they were both red,
(ii) the first ball drawn came from box A. *(WJEC)*

36 Three balls are to be drawn at random without replacement from a bag which contains four red balls, three white balls and two blue balls. Calculate the probabilities that
   (i) a red ball will not be drawn,
   (ii) one ball of each colour will be drawn,
   (iii) the three balls drawn will be of the same colour. *(WJEC)*

37 A fair coin is tossed $2n$ times.
   (i) Show that the probability that there will be at least n heads with exactly n of them occurring consecutively is
   $$\frac{(n+3)}{2^{n+2}}.$$
   (ii) The $2n$ tosses resulted in exactly n heads, not necessarily consecutive. Find, in terms of n, an expression for the probability p that both the first and last tosses were heads. Hence, given that $n > 1$, show that
   $$\frac{1}{6} < p < \frac{1}{4}.$$
   (JMB)

38 A boy has N different pairs of socks which he keeps in a drawer. He picks r socks $[0 < r < N]$ at random from the drawer. Show that the probability that they do not include a complete pair is
$$2^{r-1}\frac{(N-1)!\,(2N-r)!}{(N-r)!\,(2N-1)!}.$$

Find the probability that r socks are different but the next sock chosen completes a pair.

Given that $N > 2$ and that four socks are chosen,
   (i) show that the probability that they include at least one pair is
   $$\frac{12N-21}{(2N-1)(2N-3)}$$
   (ii) determine the probability that they comprise two pairs. *(JMB)*

39 a A game of 'passing the parcel' is played by $N + 1$ players. Initially the parcel is held by one player who passes it to another, who then passes it to a third, and so on. At each stage the next recipient is chosen at random from the N players other than the current holder. Find the probability that the parcel is passed k times
  (i) without being returned to the initial holder,
  (ii) without any player receiving it more than once.

b In a school with $N + 1$ pupils one pupil sends a letter to two others (the 'first generation') with instructions to each to copy it and send it to two others (the 'second generation'), and so on. At each stage the two recipients are chosen at random by each pupil from the N others, including the one who sent it to him. If at any time a pupil receives more than one letter he deals with each letter separately in this manner. Show that the probability that the pupil who started the 'chain letter' does not himself receive a copy before the $(k + 1)$th generation is

$$\left(1 - \frac{2}{N}\right)^{2^k - 2}.$$

(JMB)

Chapter 6

Discrete random variables

A *random variable* is a numerically valued characteristic associated with the outcomes of a random experiment, e.g. when two coins are tossed, the number of heads obtained is a random variable. More formally, a random variable is a function whose domain is the sample space of the random experiment and whose range is a set of real numbers; the range of a random variable is called its *range space*.

For example, if X is the number of heads obtained when two coins are tossed then the random variable X is the function with domain $\{HH, HT, TH, TT\}$ and range $\{0, 1, 2\}$ defined by the mapping diagram below.

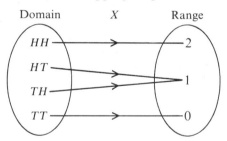

The range space of X is $\{0, 1, 2\}$.

If the number of elements in the range space is finite or countably infinite (i.e. the elements may be placed in 1–1 correspondence with the natural numbers), the random variable is called a *discrete random variable*. A random variable will be denoted by a capital letter such as X or Y; a particular numerical value of a random variable will be denoted by a lower-case letter such as x or y.

6.1 Discrete probability distributions

A function which assigns a probability to each element of the range space of a discrete random variable is called the probability function of the random variable. A more formal definition is given below.

Let X be a discrete random variable with range space $R_x = \{x_1, x_2, \ldots, x_k\}$.

Let p be a function with domain R_x such that
$$p(x_i) = P(X = x_i), \qquad\qquad i = 1, 2, \ldots, k,$$
$$p(x_i) \geqslant 0,$$
and
$$\sum_{i=1}^{k} p(x_i) = 1.$$

The function p is called the *probability function of* X. The set of ordered pairs $(x_i, p(x_i))$, $i = 1, 2, \ldots, k$, is referred to as the *probability distribution of* X.

Some possible probability functions for X, the number of heads obtained when a coin is tossed twice, are as follows:

(i) $p(0) = 0.25$, $p(1) = 0.5$, $p(2) = 0.25$;

(ii) $p(0) = 0$, $p(1) = 0$, $p(2) = 1$;

(iii) $p(0) = 0.16$, $p(1) = 0.48$, $p(2) = 0.36$.

It should be noted that, in each example,
$$p(x_i) \geqslant 0, \ (i = 1, 2, 3) \qquad \text{and} \qquad \sum_{i=1}^{3} p(x_i) = 1.$$

Only one of these probability functions can be correct for a particular coin. The first probability function would be correct if the coin is unbiased, i.e. on each toss of the coin a head or a tail is equally likely. The second would be correct if the coin is double-headed and the third probability function would be correct if the coin is biased such that $P(H) = 0.6$ on each toss. In practice the assignment of probabilities is an empirical problem, but in examination questions it follows from the description of the random experiment.

Example 1

Find the probability distribution of X, the number of heads obtained when an unbiased coin is tossed three times in succession.

$$S = \{TTT, TTH, THT, HTT, THH, HTH, HHT, HHH\}$$

$X = 0$ only when the outcome is TTT $P(X = 0) = \dfrac{1}{8}$

$X = 1$ when the outcome is TTH, THT or HTT $P(X = 1) = \dfrac{3}{8}$

$X = 2$ when the outcome is THH, HTH or HHT $P(X = 2) = \dfrac{3}{8}$

$X = 3$ only when the outcome is HHH $P(X = 3) = \dfrac{1}{8}$

The probability distribution of X is displayed in the table below, in which the notation $p(x_i) = P(X = x_i)$ is used.

x_i	0	1	2	3
$p(x_i)$	$\frac{1}{8}$	$\frac{3}{8}$	$\frac{3}{8}$	$\frac{1}{8}$

Always check that $\Sigma\, p(x_i) = 1$.

The distribution may be illustrated by means of a line diagram, as shown below.

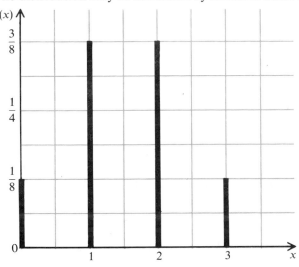

Since this diagram is symmetrical about the line $x = 1.5$ the distribution is said to be *symmetrical*.

Example 2

Find the probability distribution of Y, the number of heads in the longest run of successive heads when a fair coin is tossed three times in succession.

$S = \{TTT,\ TTH,\ THT,\ HTT,\ THH,\ HTH,\ HHT,\ HHH\}$

$Y = 0$ only when the outcome is TTT $\qquad\qquad$ $P(Y = 0) = \dfrac{1}{8}$

$Y = 1$ when the outcome is $TTH,\ THT,\ HTT$ or HTH $\qquad$ $P(Y = 1) = \dfrac{4}{8}$

$Y = 2$ when the outcome is THH or HHT $\qquad\qquad$ $P(Y = 2) = \dfrac{2}{8}$

$Y = 3$ only when the outcome is HHH $\qquad\qquad$ $P(Y = 3) = \dfrac{1}{8}$

The probability distribution of Y is displayed in the table below.

y_i	0	1	2	3
$p(y_i)$	$\frac{1}{8}$	$\frac{1}{2}$	$\frac{1}{4}$	$\frac{1}{8}$

Check that $\Sigma\, p(y_i) = 1$.

This is an example of a *skew distribution*, i.e. an unsymmetrical one, as the line diagram below shows.

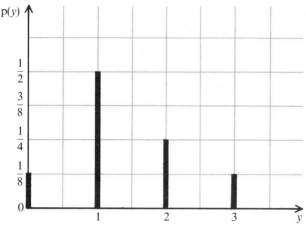

Example 3

Find the probability distribution of Z, the number of times a fair coin is tossed until a head is obtained.

$Z = 1$ when the outcome is H $\qquad$ $P(Z = 1) = \dfrac{1}{2}$

$Z = 2$ when the outcome is TH $\qquad$ $P(Z = 2) = \left(\dfrac{1}{2}\right)^2$

$Z = 3$ when the outcome is TTH $\qquad$ $P(Z = 3) = \left(\dfrac{1}{2}\right)^3$ $\qquad$ etc.

In this example, the range space of Z has an infinite number of elements and it is not possible to make a full list. However, the probability distribution of Z is given by

$$P(Z = r) = \left(\frac{1}{2}\right)^r \qquad r = 1, 2, 3, \ldots.$$

To check that $\displaystyle\sum_{r=1}^{\infty} P(Z = r) = 1$, it is necessary to observe that $\displaystyle\sum_{r=1}^{\infty} \left(\frac{1}{2}\right)^r$ is the sum of an infinite geometric series with first term $\dfrac{1}{2}$ and common ratio $\dfrac{1}{2}$.

Using the formula $S = \dfrac{a}{(1 - r)}$ (see **Result 1** on page 68) it follows that

$$\sum_{r=1}^{\infty} \left(\frac{1}{2}\right)^r = \frac{\dfrac{1}{2}}{\left(1 - \dfrac{1}{2}\right)} = 1.$$

Exercise 6.1

Find the probability distribution of X in each of the following questions.

1 Three balls are drawn at random without replacement from a bag containing five red balls and three blue balls. X is the number of red balls drawn.

2 Four unbiased coins are tossed simultaneously. X is the absolute value of the difference between the number of heads and the number of tails obtained.

3 Two fair dice are thrown. X is the smaller of the two scores on the dice.

4 Balls are drawn at random without replacement from a bag containing five red balls and three blue balls until a red ball is drawn. X is the number of balls drawn.

5 Cards are dealt at random with replacement from an ordinary pack of 52 playing-cards. X is the number of cards dealt until a heart is obtained.

6.2 Expected value of a discrete random variable

In Chapter 2, a number of useful summary measures were defined for frequency distributions of sets of data. Similar measures may be defined for probability distributions of random variables.

Suppose that X is a discrete random variable with range space $(x_1, x_2, \ldots, x_k)$ and

$$p_i = P(X = x_i), \qquad i = 1, 2, \ldots, k.$$

If in n trials of the random experiment associated with X the value x_i is observed to occur f_i times, then the mean $\bar{x}$ of the observed values is given by

$$\bar{x} = \frac{\sum\limits_{i=1}^{k} f_i x_i}{n} \qquad \left(\text{where } n = \sum\limits_{i=1}^{k} f_i \right)$$

which may be written as $\qquad \bar{x} = \sum\limits_{i=1}^{k} \frac{f_i}{n} x_i.$

Since $\dfrac{f_i}{n}$ is the relative frequency of the value x_i in the n trials, it should be approximately equal to p_i when n is large. Therefore,

for large n, $\bar{x} \approx \sum\limits_{i=1}^{k} p_i x_i.$

This motivates the following definition.

The *mean of the probability distribution of* X, denoted by μ the lower case Greek letter 'mu', is given by

$$\mu = \sum_{i=1}^{k} p_i x_i.$$

Expected value

If X is a discrete random variable with range space $\{x_1, x_2, \ldots, x_k\}$ and

$$p_i = P(X = x_i), \qquad i = 1, 2, \ldots, k,$$

then the *expected value of* X (or *the expectation of* X), denoted by $E[X]$, is defined to be

$$E[X] = \sum_{i=1}^{k} p_i x_i.$$

It follows that $\quad E[X] = \mu.$

$E[X]$ may be interpreted as the average value of X over an indefinitely large number of trials of the random experiment with which X is associated.

Example 4

Find the expected value of the score when an unbiased die is thrown once.

The probability distribution of the score is given by the first two columns of the following table and the product of the entries in these columns is shown in the third column.

x_i	p_i	$p_i x_i$
1	1/6	1/6
2	1/6	2/6
3	1/6	3/6
4	1/6	4/6
5	1/6	5/6
6	1/6	6/6

Check that $\Sigma\, p_i = 1$ $\qquad\qquad$ $6/6$ $\qquad$ $21/6 \;=\; \Sigma\, p_i x_i$

$$E[X] = \Sigma\, p_i x_i$$

$$= 21/6 = 3.5$$

Note that this distribution is symmetrical about its mean.

Example 5

Three balls are drawn at random without replacement from a bag containing four red balls and three white balls. If X is the number of red balls drawn, find the probability distribution and the mean of X.

$$P(X = 0) = P(WWW \text{ in any order}) = \frac{3}{7} \times \frac{2}{6} \times \frac{1}{5} \qquad = \frac{1}{35}$$

$$P(X = 1) = P(RWW \text{ in any order}) = \frac{4}{7} \times \frac{3}{6} \times \frac{2}{5} \times \frac{3!}{2!} = \frac{12}{35}$$

$$P(X = 2) = P(RRW \text{ in any order}) = \frac{4}{7} \times \frac{3}{6} \times \frac{3}{5} \times \frac{3!}{2!} = \frac{18}{35}$$

$$P(X = 3) = P(RRR \text{ in any order}) = \frac{4}{7} \times \frac{3}{6} \times \frac{2}{5} \qquad = \frac{4}{35}$$

The probability distribution of X is given by the first two columns in the following table.

x_i	p_i	$p_i x_i$
0	1/35	0
1	12/35	12/35
2	18/35	36/35
3	4/35	12/35

Check that $\Sigma p_i = 1$ 35/35   60/35 $= \Sigma p_i x_i$

$$\mu = E[X] = \Sigma p_i x_i$$
$$= 60/35$$

Therefore the mean of X is 12/7.

Example 6

Find the mean of the probability distribution of Z, the number of times a fair coin is tossed until a head is obtained.

As shown in **Example 3** on page 95, the probability distribution of Z is given by

$$P(Z = r) = \left(\frac{1}{2}\right)^r \qquad r = 1, 2, 3, \ldots$$

$$E[Z] = \Sigma p_i z_i$$

$$= \sum_{r=1}^{\infty} \left(\frac{1}{2}\right)^r r$$

$$= \frac{1}{2} \sum_{r=1}^{\infty} \left(\frac{1}{2}\right)^{r-1} r.$$

Using the third result in **3** on page 69, viz. $\sum_{r=1}^{\infty} rx^{r-1} = (1-x)^{-2}$ when $|x| < 1$

$$E[Z] = \frac{1}{2} \times \left(1 - \frac{1}{2}\right)^{-2}.$$

Therefore the mean of Z is 2.

Example 7

The probability distribution of the discrete random variable X is given by
$$P(X = r) = kr, \qquad r = 1, 2, \ldots, n$$
where k is a constant. Find the value of k and the mean of X.

$$\Sigma p_i = 1$$

$$\sum_{r=1}^{n} kr = 1$$

$$k \sum_{r=1}^{n} r = 1$$

Using the first result in **4** on page 69, viz. $\sum_{r=1}^{n} r = \frac{n}{2}(n + 1)$

$$\frac{kn}{2}(n + 1) = 1.$$

Therefore $\qquad k = \dfrac{2}{n(n + 1)}$

$$E[X] = \Sigma p_i x_i$$

$$= \sum_{r=1}^{n} kr \cdot r$$

$$= k \sum_{r=1}^{n} r^2.$$

Using the second result in **4** on page 69, viz. $\sum_{r=1}^{n} r^2 = \frac{n}{6}(n + 1)(2n + 1)$ and also substituting for k

$$E[X] = \frac{2}{n(n + 1)} \times \frac{n}{6}(n + 1)(2n + 1).$$

Therefore the mean of X is $\frac{1}{3}(2n + 1)$.

Exercise 6.2

Find the expected value of X in each of the questions in Exercise 6.1.

6.3 Expected value of a function

Suppose that g is a function and X is a discrete random variable with range space $(x_1, x_2, x_3, \ldots, x_k)$ such that $p_i = P(X = x_i)$, then the *expected value* (or *expectation*) of g(X) is given by

$$E[g(X)] = \sum_{i=1}^{k} p_i g(x_i).$$

For example, $E[X^2] = p_1 x_1^2 + p_2 x_2^2 + \ldots + p_k x_k^2$

$$E[t^x] = p_1 t_1^{x_1} + p_2 t_2^{x_2} + \ldots + p_k t_k^{x_k}.$$

The following results concerning the expected value operator are particularly useful. If a and b are two constants and g and h are two functions:

1        $E[b] = b$

2        $E[aX] = aE[X]$

3      $E[aX + b] = aE[X] + b$

4   $E[ag(X) + bh(X)] = aE[g(X)] + bE[h(X)].$

The proofs of these results are quite straightforward.

Consider **4**, $E[ag(X) + bh(X)] = \sum p_i(ag(x_i) + bh(x_i))$

$$= \sum (p_i ag(x_i) + p_i bh(x_i))$$

$$= \sum p_i ag(x_i) + \sum p_i bh(x_i)$$

$$= a \sum p_i g(x_i) + b \sum p_i h(x_i) \quad a, b \text{ independent of } i$$

$$= aE[g(X)] + bE[h(X)].$$

Putting g(X) = X and h(X) = 1 in this result establishes the third result; putting $b = 0$ in the third result establishes the second result; putting $a = 0$ in the third result establishes the first result. Alternatively, the four results may be proved independently.

Variance

The *variance* $V[X]$ of the probability distribution of the discrete random variable X, whose mean is μ, is defined by

$$V[X] = E[(X - \mu)^2].$$

The positive square root of $V[X]$ is called the *standard deviation* of the distribution and is denoted by $SD[X]$. Alternatively, $SD[X]$ is often denoted by σ, where σ is the lower case Greek letter 'sigma'. Both the variance and the standard deviation are used as measures of dispersion. The motivation behind the definitions is similar to that described on page 96 concerning the definition of the mean of a probability distribution.

The calculation of the value of $V[X]$ may be facilitated by the use of an alternative formula derived below.

$$V[X] = E[(X - \mu)^2]$$
$$= E[X^2 - 2\mu X + \mu^2]$$
$$= E[X^2] - 2\mu E[X] + E[\mu^2] \quad \text{(using 4)}$$

Since $E[X] = \mu$ and $E[\mu^2] = \mu^2$ it follows that

$$V[X] = E[X^2] - \mu^2$$

or, alternatively, $\quad V[X] = E[X^2] - \{E[X]\}^2.$

Example 8

Find the mean and the variance of the distribution given by

x_i	0	1	2	3
p_i	0.1	0.4	0.3	0.2

The necessary calculations to be made are displayed in the following table.

x_i	p_i	$p_i x_i$	$p_i x_i^2$
0	0.1	0	0
1	0.4	0.4	0.4
2	0.3	0.6	1.2
3	0.2	0.6	1.8
	1.0	1.6	3.4

Check that $\Sigma p_i = 1$

$$E[X] = \Sigma p_i x_i$$
$$= 1.6$$

$$V[X] = E[X^2] - \{E[X]\}^2$$
$$= \Sigma p_i x_i^2 - \{\Sigma p_i x_i\}^2$$
$$= 3.4 - 1.6^2$$
$$= 0.84$$

Exercise 6.3

Find the mean and the variance of each of the following probability distributions.

1

x_i	0	1	2	3	4
p_i	0.2	0.3	0.3	0.1	0.1

2

x_i	1	3	5	7
p_i	0.1	0.2	0.3	0.4

3

x_i	1	2	3	4
p_i	$\dfrac{1}{7}$	$\dfrac{3}{7}$	$\dfrac{2}{7}$	$\dfrac{1}{7}$

	x_i	10	20	30	40
4	p_i	$\dfrac{1}{4}$	$\dfrac{1}{4}$	$\dfrac{1}{4}$	$\dfrac{1}{4}$

	x_i	1	2	3	4	5
5	p_i	0.25	0.2	0.1	0.2	0.25

6.4 Mean and variance of a linear function

If the mean and the variance of the probability distribution of the discrete random variable X are known, it is possible to find the mean and the variance of the distribution of a linear function, Y, of X without finding the distribution of Y.

Suppose that $Y = aX + b$, where a and b are two constants.

It has already been established (on page 100) that

$$E[Y] = aE[X] + b \quad \text{or} \quad \mu_Y = a\mu_X + b.$$

By definition,
$$\begin{aligned}
V[Y] &= E[(Y - \mu_Y)^2] \\
&= E[(\{aX + b\} - \{a\mu_X + b\})^2] \\
&= E[(aX + b - a\mu_X - b)^2] \\
&= E[a^2(X - \mu_X)^2] \\
&= a^2 E[(X - \mu_X)^2] \\
&= a^2 V[X].
\end{aligned}$$

Exercise 6.4

1 The discrete random variable X has mean 3 and variance 4. Find the mean and variance of

a $X + 3$ b $2X$ c $2X + 3$

d $3X - 2$ e $3 - 2X$ f $\dfrac{(X - 3)}{2}$.

2 The discrete random variable X has mean 5 and standard deviation 2. Find the mean and standard deviation of

a $X + 1$ b $3X$ c $3X + 1$

d $4 - X$ e $4X - 3$ f $\dfrac{(X - 5)}{2}$.

3 A bag contains six red balls and four blue balls. Two balls are drawn at random without replacement from the bag, find the mean and variance of the number of red balls drawn. If 3 points are scored for each red ball and 2 points for each blue ball, deduce the mean and variance of the total score for the two balls drawn.

4 A junior salesman in an electrical shop is paid a basic wage of £50 per week together with a bonus of £2 for each television set he sells. Given that the number of television sets the salesman sells per week is a discrete random variable with mean 10 and variance 9, calculate the mean and the standard deviation of the salesman's weekly earnings.

5 The number of questions out of 100 answered correctly by candidates in a certain test is a discrete random variable with mean 20 and variance 16. Four marks are given for each correct answer and one mark is deducted for each incorrect answer. Assuming that all the candidates answer every question, find the mean and the variance of the marks scored by the candidates in the test.

6.5 Mean of a linear combination of discrete random variables

If the means of two random variables X and Y are known, it is possible to find the mean of a linear combination of X and Y without finding the probability distribution of the linear combination. In particular, the result

$$E[aX + bY] = aE[X] + bE[Y]$$

gives the mean of the random variable $(aX + bY)$ without finding its probability distribution. A proof of the above formula will be given in the chapter on joint probability distributions in Book 2.

The above formula may be easily extended to include a linear combination of three or more random variables, for example,

$$E[aX + bY + cZ] = aE[X] + bE[Y] + cE[Z].$$

Exercise 6.5

1 Given that the means of X and Y are 5 and 4 respectively, find the means of
  a $X + Y$ **b** $X - Y$ **c** $2X + Y$ **d** $3X + 4Y$.

2 Given that the means of X and Y are 3 and -2 respectively, find the means of
  a $X + Y$ **b** $X - Y$ **c** $2X + Y$ **d** $3X + 4Y$.

3 Given that the means of X, Y and Z are 3, 4 and 5 respectively, find the means of
  a $X + Y + Z$ **b** $2X + Y - Z$ **c** $X - Y + Z$.

4 A saleswoman in an electrical shop is paid a basic wage of £80 per week together with a bonus of £3 for each TV set and £1 for each radio she sells. The number of TV sets the saleswoman sells per week is a discrete random variable with mean 15, and the number of radios she sells per week is a discrete random variable with mean 5. Find the mean weekly earnings of the saleswoman.

5 The number of questions out of 100 answered correctly by candidates in a certain test is a discrete random variable with mean 50 and the number of questions answered incorrectly is a discrete random variable with mean 20. Four marks are awarded for each correct answer, one mark is deducted for each incorrect answer and no mark is awarded for an unanswered question. Find the mean mark scored by the candidates in the test.

Miscellaneous Exercise 6

1 A discrete random variable X has a probability distribution given by

$$P(X = r) = \frac{1}{5}, \qquad r = 0, 1, 2, 3, 4.$$

Find the mean and variance of X.

2 The probability distribution of a discrete random variable X is given by

$$P(X = 1) = \frac{1}{7}, \qquad P(X = 2) = \frac{2}{7}, \qquad P(X = 3) = \frac{4}{7}.$$

Find the mean and the variance of X. (*JMB*)

3 A bag contains five balls which are numbered 1, 2, 3, 4 and 5. Three balls are drawn at random without replacement from the bag. Find
  (i) the probability that the sum of the three numbers drawn is a prime number,
  (ii) the probability that the largest number drawn is 4.
Show that the expected value of the largest number drawn is 4.5. (*JMB*)

4 In a certain gambling game a player nominates an integer x from 1 to 6 inclusive and he then throws three fair cubical dice. Calculate the probabilities that the number of x's thrown will be 0, 1, 2 and 3.

The player pays 5 pence per play of the game and he receives 48 pence if the number of x's thrown is three, 15 pence if the number of x's thrown is two, 5 pence if only one x is thrown and nothing otherwise. Calculate the player's expected gain or loss per play of the game. (*JMB*)

5 The probability distribution of X, the score when a certain biased die is thrown, is given in the table below.

x	1	2	3	4	5	6
$P(X = x)$	$2k$	$3k$	$3k$	$3k$	$3k$	$4k$

Find $E\left[\dfrac{1}{X}\right]$.

6 In a game, the player tosses a coin and throws a die numbered from 1 to 6; both may be assumed to be fair. The player's score is the number on the die if the coin shows tails, and is twice this number if the coin shows heads. Find the expected value of the score, the standard deviation, and the number of possible scores which are more than two standard deviations from the expected value. (*JMB*)

7 A club with 20 members, 12 of whom are men, selects its committee by putting the 20 names in a hat and picking 4 of them at random and without replacement. Calculate (to 3 significant figures) the probabilities that:
(i) exactly three committee members are men,
(ii) both sexes are represented on the committee.
Find the average number of women on a committee chosen in this manner.   (*JMB*)

8 X is the number of heads obtained when a coin is tossed twice and the probability of a head on each toss is constant. Show that the assignment of probabilities given by

$$P(X = 0) = \frac{1}{3}, \qquad P(X = 1) = \frac{1}{3}, \qquad P(X = 2) = \frac{1}{3}$$

is not possible.

9 A door-to-door salesman calls at the door of each house in a row of five. For each house there is a probability of $\frac{3}{4}$ that he will receive a reply.

Given that there is a reply to his call there is a probability of $\frac{1}{3}$ that he will make a sale. Find the probability distribution of the number of sales and find its mean. If he makes a profit of 60p on each sale, find the expected profit resulting from his calling at the five houses.

10 A coin has probability, p, of falling heads. The coin is thrown (separate throws being independent) until a head occurs for the first time. Show that the probability that k throws are needed to attain the first head is $p(1 - p)^{k-1}$ and that the mean number of throws needed to attain the first head is $\frac{1}{p}$. (*Hint*: Compare the series obtained here with the binomial expansion of $(1 - x)^{-2}$.)   (*JMB*)

11 a Each of two packs of five cards are numbered 1 to 5. One card is selected at random from each of the two packs. Let X denote the sum of the two numbers obtained. You are paid k pence if $X = 2$ or $X = 10$, and one penny if $X = 3$ or $X = 9$, and you have to pay two pence if $X = 6$ and one penny if $X = 5$ or 7. If $X = 4$ or 8 you neither receive nor pay anything. Determine the least integer value of k so that your expected profit is greater than zero.

b You are challenged to a game in which a coin is tossed three times. You are paid six pence for each head and you pay four pence for each tail. The coin is biased so that the probability of a head is $\frac{1}{3}$. Find your expected profit or loss from a single game.   (*WJEC*)

12 Two players compete by drawing in turn and without replacement one ball at random from a box containing 4 red and 4 white balls. The winner is the player who first draws a red ball. Calculate the probability that the winner is the player who makes the first draw. Derive the corresponding probability when the balls are replaced after each draw.

In the case when the balls are not replaced calculate the mean number of balls that will be drawn in such a competition.

(*JMB*)

13 A sales representative has been assigned to a large town for one month to sell a new type of photocopying machine to business and industrial concerns in the town. Previous experience with such sales campaigns indicates that if the representative were to visit n concerns during the month then the probability that he would sell k machines is given by

$$p(k) = \frac{2}{n+1}\left(1 - \frac{k}{n}\right), \qquad k = 0, 1, 2, \ldots, n.$$

Determine the expected number of sales in terms of n.

Suppose that the representative's expenses in making n visits amount to £$\frac{1}{2}n^2$ and that he receives a commission of £60 on each machine he sells. Show that the representative's expected net profit (commission less expenses) is greatest when $n = 20$. For this value of n calculate the probabilities that at the end of the month the representative will
(i) be out of pocket,
(ii) have made a profit of at least £100.

(*WJEC*)

14 An unfair die is such that the probability of throwing a 2, 4 or 6 is p in each case, and the probability of throwing a 1, 3 or 5 is $\frac{1}{3}p$ in each case. Find p and the mean score over a large number of throws.

The unfair die is thrown with a fair die and the two scores added to form a sum X.
 (i) Find the distribution of X and show that the expected value of X is 7.25.
 (ii) Given that $X = 8$, find the conditional probability that the unfair die showed a 2.
 (iii) Find the probability that X is greater than three.

(*WJEC*)

15 A certain brand of tea has a picture card in every packet. The cards form a set of 50 different pictures and are distributed among the packets so that any packet purchased is equally likely to contain any one of the cards. A boy has collected 47 different cards. Find the probabilities that he will get the three cards he needs to complete the set
(i) if he opens only three packets,
(ii) if he opens at most four packets.

Another boy needs only one more card to complete the set. Find the probability distribution of the number of packets this boy needs to open to complete the set. Calculate to two decimal places the probability that he gets the remaining card by opening not more than ten packets.

(*JMB*)

16 A box contains 9 discs, of which 4 are red, 3 are white and 2 are blue. Three discs are to be drawn at random without replacement from the box. Find
 (i) the probability that the discs, in the order drawn, will be coloured red, white and blue, respectively,
 (ii) the probability that one disc of each colour will be drawn,
 (iii) the probability that the third disc drawn will be red,
 (iv) the probability that no red disc will be drawn,
 (v) the most probable number of red discs that will be drawn,
 (vi) the expected number of red discs that will be drawn, and state the probability that this expected number of red discs will be drawn.

(*JMB*)

17 A batch of 20 items is inspected as follows. A random sample of 5 items is drawn from the batch without replacement and the number of defective items in the sample is counted. If this number is 2 or more the batch is rejected; if there is no defective item in the sample the batch is accepted; if there is exactly one defective item in the sample, then a further random sample of 5 items is drawn from the remaining 15 items in the batch. If this second sample includes at least 1 defective item then the batch is rejected; otherwise the batch is accepted.

Suppose that a batch to be inspected consists of exactly 2 defective items and 18 non-defective items.

a Calculate the probabilities that
 (i) the batch will be accepted on the basis of the first sample,
 (ii) a second sample will be taken and the batch will then be accepted,
 (iii) the batch will be rejected.

b Find the expected number of items that will have to be sampled to reach a decision on the batch.

(*WJEC*)

18 Each trial of a random experiment must result in one and only one of the events A, B and C occurring, the probabilities of these events being $\frac{1}{2}, \frac{1}{4}$ and $\frac{1}{4}$ respectively. Independent trials of the experiment are to be conducted until one of the three events occurs for the second time.
 (i) Show that the probability that the trials will stop with A occurring for the second time
   (a) in the second trial is $\frac{1}{4}$,
   (b) in the third trial is $\frac{1}{4}$,
   (c) in the fourth trial is $\frac{3}{32}$.
 (ii) Determine the corresponding probabilities that the trials will stop with B occurring for the second time.
 (iii) Let X denote the number of trials that will be conducted. Find the distribution of X. Hence determine the most probable value of X and the expected value of X.

(*WJEC*)

19 A box contains n balls, of which two are white. Balls are to be drawn at random from the box, one after another without replacement, until both white balls have been drawn. Let X denote the number of balls that will be drawn.

Show that $P(X = r) = \dfrac{2(r-1)}{n(n-1)}$, $\qquad r = 2, 3, \ldots, n$.

Find, in as simple a form as possible, an expression for the mean value of X. *(JMB)*

20 A and B play a match consisting of a series of games. The winner of the match is the player who first wins two games more than his opponent. Experience of contests between the two has shown that the probability that A wins any game is p. Games cannot be drawn. Write down the probabilities that A wins the match

(i) after two games, (ii) after three games, (iii) after four games.

Show that the probability that A wins the match is

$$\frac{p^2}{1 - 2p + 2p^2}.$$

Find the mean number of games in a completed match. *(JMB)*

21 A game of chance consists of throwing a disc of radius 2in on to a horizontal circular board of radius 22 in marked with concentric rings of negligible thickness as shown in the diagram. The rings have radii 6in, 12in and 18in respectively and there is a rim round the edge of the board to prevent the disc falling off.

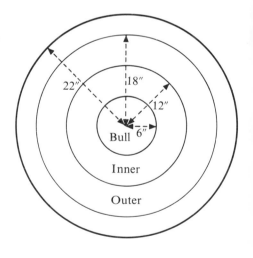

A player 'wins' if, when the disc settles, it lies entirely within one of the regions marked Bull, Inner, Outer, (i.e. it does not lie across any of the bounding rings).

Calculate the probabilities that a disc thrown at random will fall entirely within the regions marked

(i) Bull, (ii) Inner, (iii) Outer,

and deduce that the probability of a 'miss' (i.e. not a 'win') is $\dfrac{18}{25}$. [Assume that the final position of the centre of the disc is uniformly distributed over the accessible region.]

If prizes of 50p, 25p, 10p are awarded when the disc falls entirely within the regions Bull, Inner and Outer respectively, show that a charge of 6p per throw should enable the owner to make a profit in the long run.

Calculate the probabilities that a player throwing four times at random will

  (i) score exactly 3 Outers,

 (ii) win prizes to the value of exactly 30p,

(iii) win prizes to the value of exactly 35p.   (*JMB*)

22 Define the mean and variance of a discrete random variable X which has a given probability function p(x).

In a 'roll-a-penny' game at a fairground a penny is rolled onto a table marked in squares of equal size. The length of the sides of the squares is twice the diameter of the penny. If the penny lies fully within a square it is returned to the player. The centre of the penny may be assumed to lie at random on the marked part of the table, and the lines marking the squares are of negligible thickness. Determine the probability that the penny is returned to the player.

For some squares an additional 'prize' of threepence or sixpence is given. If one third of the squares yield a prize of threepence, and one ninth of the squares yield a prize of sixpence, calculate the mean and variance of the profit to the player which arises from rolling a single penny.   (*JMB*)

Chapter 7

Special discrete distributions

In many random experiments, the only point of interest in each trial is the occurrence or non-occurrence of a particular event; the occurrence of the event is called a 'success' and the non-occurrence of the event is called a 'failure'. Such a trial is called a Bernoulli trial, after Jacques Bernoulli (1654–1705), also known as Jakob or James, who discovered the binomial distribution. It should be noted that the words 'success' and 'failure' used in this sense do not have their usual connotations. The tossing of a coin is a simple example of a Bernoulli trial since for each toss there are two possible outcomes, either of which may be considered to be a 'success' and the other a 'failure'. If, when a die is thrown, the score of 6 is deemed to be a 'success' and any other score a 'failure', the throwing of the die is another example of a Bernoulli trial.

7.1 Binomial distribution

If X is the number of successes in n independent Bernoulli trials, in each of which the probability of a success is a constant p and the probability of a failure is $q = 1 - p$, then

$$P(X = r) = \binom{n}{r} q^{n-r} p^r, \qquad r = 0, 1, 2, \ldots, n.$$

The distribution of X is called the binomial distribution with index n and probability parameter p; this is often abbreviated to $X \sim B(n, p)$.

Proof

Since the trials are independent and p, q are constants, the probability of $(n - r)$ failures followed by r successes is given by

$$P(\overset{\leftarrow (n-r) \rightarrow}{FF \ldots F} \overset{\leftarrow \ r \ \rightarrow}{SS \ldots S} \text{ in order}) = q^{n-r} p^r.$$

Since the symbols $FF \ldots F SS \ldots S$ may be arranged in

$$\frac{n!}{(n-r)! \, r!}$$

ways, the probability of $(n - r)$ failures and r successes in any order is given by

$$P(FF\ldots F\, SS\ldots S \text{ in any order}) = \frac{n!}{(n-r)!\,r!} q^{n-r}p^r;$$

therefore
$$P(X = r) = \binom{n}{r} q^{n-r}p^r, \qquad r = 0, 1, 2, \ldots, n.$$

The binomial distribution is so named because $P(X = r)$ is the coefficient of t^r in the binomial expansion of $(q + pt)^n$. It is an important distribution as it may be used as a model in a wide range of situations. Some examples of random variables which may be regarded as binomially distributed (or approximately so) are given below.

(i) The number of defective items in a sample of size n taken from a large batch of mass-produced items as part of a quality control process.

(ii) The number of seeds germinating when n seeds in a packet are planted.

(iii) The number of sixes obtained when a die is thrown n times.

(iv) In a poll of n persons, the number who say they will vote for a particular party.

In the above examples, it is assumed that, in each trial of the experiment, the probability of a success is constant (or approximately so) and that the trials are independent.

Example 1

A psychologist wishes to study people who claim that they have extra-sensory perception. Initially he uses the following procedure to test their claims. The psychologist deals a card at random from a standard pack and notes the colour (red or black) of the card; the subject is then asked to write down the colour of the card. The experiment is repeated ten times and the number of times the subject has correctly matched the cards is noted; the subject passes the test if at least eight matches are obtained. Calculate the probability of passing the test by guessing.

Let X be the number of correct matches obtained in 10 trials by guessing.

$$X \sim B\left(10, \frac{1}{2}\right)$$

$$P(X = 8) = \binom{10}{8}\left(\frac{1}{2}\right)^2\left(\frac{1}{2}\right)^8 = \frac{45}{1024}$$

$$P(X = 9) = \binom{10}{9}\left(\frac{1}{2}\right)\left(\frac{1}{2}\right)^9 = \frac{10}{1024}$$

$$P(X = 10) = \binom{10}{10}\left(\frac{1}{2}\right)^0\left(\frac{1}{2}\right)^{10} = \frac{1}{1024}$$

$$P(X \geqslant 8) = \frac{45 + 10 + 1}{1024} = \frac{7}{128}$$

The calculations in **Example 1** were easily performed without a calculator. In questions in which the calculations are more difficult, the following recursive formula may be useful.

$$\frac{p_r}{p_{r-1}} = \frac{(n-r+1)p}{rq}$$

where $p_r = P(X = r)$ and $p_{r-1} = P(X = r - 1)$

Proof $$\frac{p_r}{p_{r-1}} = \binom{n}{r} q^{n-r} p^r \div \binom{n}{r-1} q^{n-r+1} p^{r-1}$$

$$\frac{p_r}{p_{r-1}} = \frac{n!\, q^{n-r} p^r}{r!\,(n-r)!} \times \frac{(r-1)!\,(n-r+1)!}{n!\, q^{n-r+1} p^{r-1}}$$

$$\frac{p_r}{p_{r-1}} = \frac{(n-r+1)p}{rq}$$

Example 2

A factory produces silicon chips in very large batches. Given that 15% of the silicon chips in a batch are defective, find the probability that a random sample of 50 of these silicon chips will contain fewer than 5 defective chips.

Since the number in the sample is very small compared with the number of chips in the batch, the probability of choosing a defective one on each occasion may be considered to remain constant at 0.15.

Let X be the number of defectives in the random sample of 50 chips.

$$X \sim B(50, 0.15)$$

$$p_0 = (0.85)^{50} = 0.000295764$$

Put this into the memory. Do not clear the display. Use the recursive formula with $n = 50$, $r = 1$, $p = 0.15$ and $q = 0.85$.

$$p_1 = p_0 \times \frac{50 \times 0.15}{1 \times 0.85} = 0.00260968$$

Add this to the memory. Do not clear the display. Put $r = 2$ in the formula.

$$p_2 = p_1 \times \frac{49 \times 0.15}{2 \times 0.85} = 0.011283$$

Add this to the memory. Do not clear the display. Put $r = 3$ in the formula.

$$p_3 = p_2 \times \frac{48 \times 0.15}{3 \times 0.85} = 0.031858$$

Add this to the memory. Do not clear the display. Put $r = 4$ in the formula.

$$p_4 = p_3 \times \frac{47 \times 0.15}{4 \times 0.85} = 0.0660586$$

Add this to the memory. Recall the memory to give the required answer.

Therefore $P(X < 5) = 0.1121$ (to 4 d.p.).

Calculations such as those in **Example 2** are tedious to perform on a calculator but in appropriate cases they may be greatly simplified by using tables.

Use of tables

Tables of cumulative binomial probabilities exist in a number of different forms. Some list $P(X \leqslant r)$ for various values of n, p and r; others list $P(X \geqslant r)$. If such tables are available in an examination they may be used when the values of n, p and r in the question appear in the table. **Table 1** on page 215 which lists $P(X \leqslant r)$ for certain values of n, p and r, should be used in the following examples.

Example 3

Given that $X \sim B(20, 0.2)$, find   **a** $P(X < 5)$ **b** $P(X \geqslant 2)$ **c** $P(X = 7)$.

a $P(X < 5) = P(X \leqslant 4) = 0.630$

b $P(X \geqslant 2) = 1 - P(X \leqslant 1) = 1 - 0.069 = 0.931$

c $P(X = 7) = P(X \leqslant 7) - P(X \leqslant 6) = 0.968 - 0.913 = 0.055$

Example 4

Given that $X \sim B(10, 0.6)$, find   **a** $P(X = 2)$ **b** $P(3 < X < 7)$.

The value $p = 0.6$ is not tabulated, so consider Y, the number of failures.

$Y \sim B(10, 0.4)$

a $P(X = 2) = P(Y = 8) = P(Y \leqslant 8) - P(Y \leqslant 7) = 0.998 - 0.988 = 0.010$

b $P(3 < X < 7) = P(4 \leqslant Y \leqslant 6) = P(Y \leqslant 6) - P(Y \leqslant 3) = 0.945 - 0.382 = 0.563$

Exercise 7.1

1 $X \sim B\left(12, \frac{1}{3}\right)$, find   **a** $P(X \leqslant 4)$ **b** $P(X = 4)$.

2 $X \sim B(30, 0.2)$, find   **a** $P(X \geqslant 3)$ **b** $P(X = 5)$.

3 $X \sim B(20, 0.3)$, find   **a** $P(X \leqslant 4)$ **b** $P(X = 2)$.

4 $X \sim B(20, 0.9)$, find   **a** $P(X > 16)$ **b** $P(X = 18)$.

5 $X \sim B\left(15, \frac{1}{7}\right)$, find   **a** $P(X \geqslant 2)$ **b** $P(X = 2)$.

6 $X \sim B(50, 0.7)$, find   **a** $P(X \leqslant 37)$ **b** $P(X = 30)$.

7 $X \sim B(10, 0.45)$, find   **a** $P(X \geqslant 3)$ **b** $P(X = 5)$.

8 $X \sim B(10, 0.65)$, find   **a** $P(X > 4)$ **b** $P(5 < X < 9)$.

9 $X \sim B(5, 0.05)$, find   **a** $P(X < 2)$ **b** $P(X = 0)$.

10 $X \sim B(5, 0.75)$, find   **a** $P(X < 3)$ **b** $P(X = 6)$.

7.2 Probability generating functions

If X is a discrete random variable whose range space R_x is a subset of W, where $W = \{\text{whole numbers}\}$, then the function G defined by

$$G(t) = \sum_{r=0}^{\infty} p_r t^r$$

where $p_r = P(X = r)$, is called the *probability generating function* of the distribution of X. It should be noted that the coefficient of t^r is $P(X = r)$ and that t is a 'dummy' variable, having no significance in itself.

An alternative equivalent definition is $G(t) = E[t^X]$.

One immediate consequence of the definitions is that $G(1) = 1$,

since $$G(1) = \sum_{r=0}^{\infty} p_r = 1 \text{ (probability of a certain event)},$$

or, alternatively, $G(1) = E[1^X] = E[1] = 1$.

It is possible to find an expression for $E[X]$ in terms of the probability generating function of X.

In general, $$E[X] = \sum_{i=1}^{k} p_i x_i.$$

When the range space of X is W, this may be written

$$E[X] = \sum_{r=0}^{\infty} p_r r,$$

since $x_1 = 0$, $x_2 = 1$, $x_3 = 2$, etc.

Now $$G(t) = \sum_{r=0}^{\infty} p_r t^r.$$

Differentiating with respect to t $$G'(t) = \sum_{r=0}^{\infty} p_r r t^{r-1},$$

and putting $t = 1$ $$G'(1) = \sum_{r=0}^{\infty} p_r r.$$

Therefore $E[X] = G'(1)$ or $\mu = G'(1)$.   (1)

It is also possible to find an expression for the variance of the distribution in terms of its probability generating function.

From the above $$G'(t) = \sum_{r=0}^{\infty} p_r r t^{r-1}.$$

Differentiating with respect to t $\qquad G''(t) = \sum_{r=0}^{\infty} p_r r(r-1)t^{r-2},$

and putting $t = 1$ $\qquad G''(1) = \sum_{r=0}^{\infty} p_r r(r-1)$

$$G''(1) = \sum_{r=0}^{\infty} p_r r^2 - \sum_{r=0}^{\infty} p_r r$$

$$G''(1) = E[X^2] - E[X].$$

Rearranging $\qquad E[X^2] = G''(1) + \mu.$

Since $\qquad V[X] = E[X^2] - \{E[X]\}^2$

substituting for $E[X^2]$, $\qquad V[X] = G''(1) + \mu - \mu^2.$ $\qquad$ (2)

The two formulae (1) and (2) above are very useful for determining the mean and the variance of distributions whose probability generating functions have simple forms.

Mean and variance of the binomial distribution

The probability generating function of the binomial distribution $B(n, p)$ is given by

$$G(t) = (q + pt)^n$$

Proof $\qquad G(t) = \sum_{r=0}^{n} P(X = r)t^r$

$$G(t) = \sum_{r=0}^{n} \binom{n}{r} q^{n-r} p^r t^r$$

$$G(t) = \sum_{r=0}^{n} \binom{n}{r} q^{n-r} (pt)^r$$

Using the binomial theorem $\qquad G(t) = (q + pt)^n$

Differentiating with respect to t $\qquad G'(t) = n(q + pt)^{n-1} \times p$

Putting $t = 1$ $\qquad G'(1) = np(q + p)^{n-1}$

Since $q + p = 1$ $\qquad G'(1) = np$

Using (1) $\qquad \mu = np$

Differentiating $G'(t)$ $\qquad G''(t) = n(n-1)(q + pt)^{n-2} \times p^2$

Putting $t = 1$ $\qquad G''(1) = n(n-1)(q + p)^{n-2} \times p^2$

Since $q + p = 1$ $\qquad G''(1) = n(n-1)p^2$

Using (2) $\qquad V[X] = n(n-1)p^2 + np - (np)^2$

$$V[X] = n^2 p^2 - np^2 + np - n^2 p^2$$

$$V[X] = np(1 - p)$$

Hence, the mean and variance of $B(n, p)$ are np and npq, respectively.

Example 5

The random variable X is binomially distributed with mean 6.5 and variance 2.275. Find the most probable value of X.

Since the mean is 6.5 $\qquad\qquad np = 6.5.$

Since the variance is 2.275 $\qquad npq = 2.275.$

Dividing $\qquad\qquad\qquad\qquad q = \dfrac{2.275}{6.5} = 0.35.$

Since $p = 1 - q$ $\qquad\qquad p = 1 - 0.35 = 0.65.$

Substituting $\qquad\qquad\qquad n = 10.$

Using the recursive formula $\qquad \dfrac{p_r}{p_{r-1}} = \dfrac{(n - r + 1)p}{rq}$

$$\dfrac{p_r}{p_{r-1}} = \dfrac{(11 - r)0.65}{0.35r}.$$

$p_r > p_{r-1}$ if and only if $\quad (11 - r)0.65 > 0.35r$

$$7.15 - 0.65r > 0.35r$$
$$7.15 > r.$$

Therefore $p_7 > p_6 > p_5 > \ldots > p_0$

$p_r < p_{r-1}$ if and only if $\quad (11 - r)0.65 < 0.35r$

$$7.15 < r.$$

Therefore $p_{10} < p_9 < p_8 < p_7$

Therefore the most probable value of X is 7. This is also called the *mode* of the distribution.

Example 6

When a particular type of flower seed is planted, the probability that a yellow flower will result is $\frac{1}{3}$. If 243 rows of 5 seeds are planted find the expected frequency distribution of the number of yellow flowers per row.

Let X be the number of yellow flowers per row.

$$X \sim B\left(5, \frac{1}{3}\right)$$

$$P(X = 0) = \binom{5}{0}\left(\frac{2}{3}\right)^5\left(\frac{1}{3}\right)^0 = \frac{32}{243}$$

$$P(X = 1) = \binom{5}{1}\left(\frac{2}{3}\right)^4\left(\frac{1}{3}\right) = \frac{80}{243}$$

$$P(X = 2) = \binom{5}{2}\left(\frac{2}{3}\right)^3\left(\frac{1}{3}\right)^2 = \frac{80}{243}$$

$$P(X = 3) = \binom{5}{3}\left(\frac{2}{3}\right)^2\left(\frac{1}{3}\right)^3 = \frac{40}{243}$$

$$P(X = 4) = \binom{5}{4}\left(\frac{2}{3}\right)\left(\frac{1}{3}\right)^4 = \frac{10}{243}$$

$$P(X = 5) = \binom{5}{5}\left(\frac{2}{3}\right)^0\left(\frac{1}{3}\right)^5 = \frac{1}{243}$$

Let Y_i be the number of rows out of 243 containing i yellow flowers.

$$Y_0 \sim B\left(243, \frac{32}{243}\right)$$

$$E[Y_0] = np = 243 \times \frac{32}{243} = 32$$

Similarly the expected values of Y_1, Y_2, Y_3, Y_4 and Y_5 are 80, 80, 40, 10 and 1. Thus, the required expected frequency distribution of the number of yellow flowers per row is given by the following table.

No. of yellow flowers	0	1	2	3	4	5
Expected frequency	32	80	80	40	10	1

Exercise 7.2

1 Given that X is binomially distributed with mean 6 and variance 4.2, find the most probable value of X.

2 Find the mean and the mode of $B(32, 0.7)$.

3 Three fair coins are tossed simultaneously 32 times. Find the expected frequency distribution of the number of heads per toss.

4 Four fair dice are thrown simultaneously 1296 times. Find the expected frequency distribution of the number of sixes per throw.

5 A salesman calls at the door of 100 houses each week. For each house there is a probability of $\frac{1}{4}$ that he does not receive a reply. When there is a reply, the probability that he will make one sale at the house is $\frac{1}{3}$. Given that the salesman makes at most one sale per house identify the distribution of the number of sales made per week and write down its mean and variance.

Given that he is paid a basic weekly wage of £60 and a commission of £2 for each sale, find the mean and the variance of the salesman's weekly earnings.

7.3 Other distributions

Discrete uniform distribution

If a discrete random variable X with a finite range space has a probability function p, such that $p(x) = k$, where k is a constant, then X is said to have a discrete uniform distribution. The number uppermost when an unbiased die is thrown once is an example of a random variable which has a discrete uniform distribution.

Example 7

The distribution of a discrete random variable X is given by

$$P(X = r) = k, \qquad r = 1, 2, 3, \ldots, n.$$

Find the value of k and the mean and variance of X.

Since $\qquad \Sigma p_i = 1$

$$\sum_{r=1}^{n} k = 1$$

$$nk = 1$$

$$\therefore \quad k = \frac{1}{n}$$

$$E[X] = \Sigma p_i x_i$$

$$= \sum_{r=1}^{n} kr$$

$$= k \sum_{r=1}^{n} r$$

Using the first result in **4** on page 69, viz. $\sum\limits_{r=1}^{n} r = \frac{n}{2}(n + 1)$, and also substituting for k.

$$E[X] = \frac{1}{n} \times \frac{n}{2}(n + 1)$$

Therefore the mean of X is $\frac{1}{2}(n + 1)$.

$$E[X^2] = \Sigma p_i x_i^2$$

$$= \sum_{r=1}^{n} kr^2$$

$$= k \sum_{r=1}^{n} r^2$$

Using the second result in **4** on page 69, viz. $\sum\limits_{r=1}^{n} r^2 = \frac{n}{6}(n + 1)(2n + 1)$, and also substituting for k.

$$E[X^2] = \frac{1}{n} \times \frac{n}{6}(n + 1)(2n + 1)$$

$$= \frac{1}{6}(n + 1)(2n + 1)$$

$$V[X] = E[X^2] - \{E[X]\}^2$$

$$= \frac{1}{6}(n + 1)(2n + 1) - \frac{(n + 1)^2}{4}$$

Simplifying, the variance of X is $\frac{1}{12}(n + 1)(n - 1)$.

Mean and variance of a proportion

In each of n independent Bernoulli trials the probability of a success is a constant p. If X is the number of successes in the n trials, the proportion $\hat{P}$ of successes obtained is given by

$$\hat{P} = \frac{X}{n}.$$

$$\begin{aligned}
E[\hat{P}] &= E\left[\frac{X}{n}\right] \\
&= \frac{E[X]}{n} \\
&= \frac{np}{n} \\
&= p
\end{aligned}$$

$$\begin{aligned}
V[\hat{P}] &= V\left[\frac{X}{n}\right] \\
&= \frac{V[X]}{n^2} \\
&= \frac{np(1-p)}{n^2} \\
&= \frac{p(1-p)}{n}.
\end{aligned}$$

When the probability parameter p is unknown, the value of $\hat{P}$ obtained from a series of Bernoulli trials is used as an estimate for the value of p. The proportion $\hat{P}$ is called an *unbiased estimator* for the population proportion p, because its expected value is p, the value being estimated. In this situation the standard deviation of $\hat{P}$ is sometimes called the *standard error* of $\hat{P}$. Estimation will be discussed in more detail in Book 2.

Example 8

A seed merchant produces large numbers of a certain type of flower seed. If 92 seeds from a random sample of 100 seeds germinate, estimate the proportion of all the seeds produced by the merchant that will germinate. Also calculate an approximate value for the standard error of this estimate.

Estimated proportion $\hat{p} = \dfrac{92}{100} = 0.92$

Standard error $= \sqrt{\dfrac{p(1-p)}{n}}$

Estimated standard error $= \sqrt{\dfrac{\hat{p}(1 - \hat{p})}{n}}$

$$= \sqrt{\dfrac{0.92(1 - 0.92)}{100}}$$

$$= 0.0271 \text{ (4 d.p.)}$$

Geometric distribution

When a sequence of independent Bernoulli trials, in each of which the probability of a success is a constant p and the probability of a failure is $q = (1 - p)$, is carried out until a success occurs (e.g. a die is thrown until a six appears), the distribution of the number of trials carried out is said to have a geometric distribution. If X denotes the number of trials carried out, then $X = r$ when a sequence of $(r - 1)$ failures is followed by a success. Thus $P(X = r) = q^{r-1}p$.

A discrete random variable X which has a distribution given by

$$P(X = r) = pq^{r-1}, \qquad r = 1, 2, 3, \ldots$$

where $q = 1 - p$, is said to have the *geometric distribution* with probability parameter p.

Although this distribution does not appear explicitly in many A level syllabuses, particular cases (such as **Example 6** in Chapter 6) are sometimes set in examination questions.

Example 9

Find the probability generating function of the geometric distribution with probability parameter p and deduce the value of its mean.

The probability generating function of the geometric distribution is given by

$$G(t) = \sum_{r=1}^{\infty} P(X = r)t^r$$

$$= \sum_{r=1}^{\infty} pq^{r-1}t^r$$

$$= pt \sum_{r=1}^{\infty} (qt)^{r-1}$$

Using the second result in **3** on page 69, viz. $\displaystyle\sum_{r=0}^{\infty} x^r = (1 - x)^{-1}$.

$$G(t) = pt(1 - qt)^{-1}, \text{ provided that } |qt| < 1.$$

Differentiating w.r.t. t $\qquad G'(t) = p(1 - qt)^{-1} + pt \cdot q(1 - qt)^{-2}$

Putting $t = 1$ $\qquad G'(1) = p(1 - q)^{-1} + pq(1 - q)^{-2}$

Since $p = 1 - q$

$$G'(1) = 1 + \frac{q}{p} = \frac{p + q}{p} = \frac{1}{p}$$

Therefore

$$\mu = \frac{1}{p}$$

It is left as an exercise for the reader to prove that the variance of the geometric distribution is $\frac{q}{p^2}$.

Hypergeometric distribution

Suppose that in a set of N objects a proportion p have a certain characteristic and a proportion q, where $q = 1 - p$, do not have the characteristic. If X denotes the number of objects possessing the characteristic in a set of n objects chosen at random without replacement from the N objects then

$$P(X = r) = \frac{\binom{Np}{r}\binom{Nq}{n-r}}{\binom{N}{n}}.$$

X is said to have a *hypergeometric distribution*.

Proof

The number of ways in which r objects may be selected from the Np objects having the characteristic is $\binom{Np}{r}$.

The number of ways in which $(n - r)$ objects may be selected from the Nq objects not having the characteristic is $\binom{Nq}{n-r}$.

Thus the number of ways in which a set of n objects, exactly r of which have the characteristic, may be selected from the N objects is $\binom{Np}{r} \times \binom{Nq}{n-r}$.

The number of ways in which n objects may be selected from the N objects without restriction is $\binom{N}{n}$.

$$\therefore \quad P(X = r) = \frac{\binom{Np}{r} \times \binom{Nq}{n-r}}{\binom{N}{n}}$$

Example 10

Four balls are drawn at random without replacement from a box containing 3 red and 7 blue balls. Find the mean and variance of the number of red balls drawn.

$$P(X = 0) = P(BBBB) \qquad\qquad = \frac{7}{10} \times \frac{6}{9} \times \frac{5}{8} \times \frac{4}{7} \qquad\qquad = \frac{5}{30}$$

$$P(X = 1) = P(BBBR \text{ in any order}) = \frac{7}{10} \times \frac{6}{9} \times \frac{5}{8} \times \frac{3}{7} \times \frac{4!}{3!} = \frac{15}{30}$$

$$P(X = 2) = P(BBRR \text{ in any order}) = \frac{7}{10} \times \frac{6}{9} \times \frac{3}{8} \times \frac{2}{7} \times \frac{4!}{2!\,2!} = \frac{9}{30}$$

$$P(X = 3) = P(BRRR \text{ in any order}) = \frac{7}{10} \times \frac{3}{9} \times \frac{2}{8} \times \frac{1}{7} \times \frac{4!}{3!} = \frac{1}{30}$$

x_i	p_i	$p_i x_i$	$p_i x_i^2$
0	5/30	0	0
1	15/30	15/30	15/30
2	9/30	18/30	36/30
3	1/30	3/30	9/30

Check that $\Sigma p_i = 1$ 30/30   36/30   60/30

$$E[X] = 36/30 = 1.2$$

$$V[X] = E[X^2] - \{E[X]\}^2$$

$$= 60/30 - 1.2^2$$

$$= 0.56$$

The distribution of X is a particular example of a hypergeometric distribution. In general it may be proved that

$$E[X] = np$$

and $\qquad V[X] = \dfrac{(N - n)npq}{(N - 1)}.$

Also when N is large in comparison with n, it may be shown that the hypergeometric distribution may be approximated by the binomial distribution. The proofs of these results are involved and will not be given here.

In a situation in which random sampling without replacement is carried out, the hypergeometric distribution may be used as a model for the distribution of the number of items in the sample possessing a certain characteristic. In practice, when the sample size is small in comparison with that of the population, the more convenient binomial approximation is used.

Exercise 7.3

1 The distribution of a discrete random variable X is given by

$$P(X = r) = k, \qquad r = 0, 1, 2, \ldots, n.$$

Find the mean and variance of X.

2 Four hundred voters from a certain constituency were selected at random and asked about their voting intentions in a forthcoming election. If 168 stated that they would vote for a particular party, estimate the proportion of the electorate in the constituency that will vote for the party in the election. Calculate an approximate value for the standard error of your estimate.

3 A committee of a club consists of 4 men and 5 women. A sub-committee of three is to be selected at random from the committee members. Find the mean and variance of the number of women selected for the sub-committee. Deduce the mean and variance of the number of men selected for the sub-committee.

4 A box contains four white balls and two black balls. Three players A, B and C each draw in turn, with replacement, one ball at random from the box, A starting. The first to draw a black ball wins. Find the probabilities of winning for A, B and C.

If X is the random variable which takes the value r if the game finishes on the rth draw, show that the probability generating function of X is given by

$$G(t) = \frac{t}{3 - 2t}$$

and hence evaluate $E[X]$ and $V[X]$.

In the case when the balls are not replaced after drawing, find the probabilities of winning for A, B and C. *(JMB)*

5 In the game of 'odd one out', each of $n \geqslant 3$ persons simultaneously tosses a fair coin. If one coin turns up differently from the other $(n - 1)$ coins, then the person who tossed that coin is the 'odd one out'. Find the probability distribution of the number of times the coins have to be tossed for there to be an 'odd one out', and show that the mean number is equal to

$$\frac{2^{n-1}}{n}.$$

(JMB)

Miscellaneous Exercise 7

1 Given that X is binomially distributed with mean 4 and variance 3.2, find the most probable value of X.

2 Find the mode of $B(15, 0.4)$.

3 A computer installation has 10 terminals. Independently, the probability that any one particular terminal will require attention during a week is 0.1. Use tables to find the probabilities that

(i) 0,   (ii) 1,   (iii) 2,   (iv) 3 or more terminals

will require attention during the next week.

4 The probability that a certain type of seed will germinate is 0.75. Use tables to find the probabilities that, when the 5 seeds in a packet are planted, the number of seeds germinating will be
(i) 0, (ii) 1, (iii) 2, (iv) 3, (v) 4, (vi) 5.

If the seeds from 100 packets, each containing 5 seeds, are planted, find the expected frequency distribution of the number of seeds germinating per packet, giving your answers to the nearest integer.

5 The efficiency of electronic chips is checked by examining samples of five. The frequency distribution of the number of defective chips per sample obtained when 100 samples have been examined is given below.

No. of defectives	0	1	2	3	4	5
No. of samples	47	34	16	3	0	0

Calculate the proportion of defective chips in the 500 chips tested. Assuming that the binomial distribution holds, use this value to calculate the expected frequencies corresponding to the observed frequencies in the table.

6 To test whether a damaged coin is unbiased it is decided to toss the coin 20 times and to judge it as unbiased only if 8, 9, 10, 11 or 12 heads are tossed.
Use tables to calculate the probability that it will be judged
 (i) biased when, in fact, it is unbiased,
(ii) unbiased when, in fact, the probability of a head is 0.45 on each toss.

7 The random variable X has a binomial distribution with mean 7 and variance 4.55. Use tables to find the probability that X is at least 5.

(*JMB*)

8 In an agricultural experiment twelve seeds of a particular variety of radish were planted in each of 200 specially prepared plots of soil, and the number of seeds which germinated was noted for each plot. The following results were obtained.

No. of seeds germinating	0	1	2	3	4	5	6	7	8	9	10	11	12
Frequency		0	0	0	1	0	1	1	4	18	38	64	54	19

Represent this distribution graphically, and determine the sample mean and variance of the number of seeds which germinate. Estimate what proportion of seeds germinate, on average.

State, with reasons, what standard form of distribution would be expected to arise from such an experiment. (*JMB*)

9 A market gardener sowed 20 seeds of a particular plant in each of 100 specially prepared trays. The number of seeds that germinated per tray had the following frequency distribution.

Number germinating	20	19	18	17	16	15	14 or fewer
Number of trays	73	15	5	4	2	1	0

(i) Exhibit the distribution diagrammatically.

(ii) Calculate the mean and standard deviation of the distribution.

(iii) Calculate the overall proportion of seeds that germinated.

(iv) State, with reasons, what theoretical distribution could be expected to apply in the above situation. *(WJEC)*

10 A multiple-choice test paper consists of ten questions. In each question, the candidate has to choose the correct answer from a list of five answers. Consider a candidate who chooses the answer to each question at random from the listed answers. What is the probability distribution of the number of correct answers for such a candidate and what is his expected number of correct answers?

Suppose each correct answer is awarded 4 marks and each incorrect answer carries a penalty of 1 mark. Determine the expected overall mark of a candidate who randomly chooses the answer to each question.

Suppose a candidate has probability $\frac{1}{4}$ of knowing the correct answer to each question, and that when he does not know the correct answer he chooses the answer randomly. For this candidate, determine the expected number of correct answers and his expected mark on the paper. *(WJEC)*

11 Define the binomial distribution, stating the conditions under which it will arise.

A biased cubical die has a probability p of yielding an even score when thrown. What is the probability that no even scores are obtained in ten independent throws of the die?

In ten independent throws of the die the probability of exactly five even scores is twice the probability of exactly four even scores. Show that $p = \frac{5}{8}$ and calculate (to 3 places of decimals) the probability of at most six even scores in eight independent throws of the die. *(JMB)*

12 In a large school, 80% of the pupils like mathematics. A visitor to the school asks each of 4 pupils, chosen at random, whether he likes mathematics. Calculate the probabilities of obtaining the answer 'yes' from 0, 1, 2, 3, 4 pupils.

Find the probability that the visitor obtains the answer 'yes' from at least 2 pupils

(i) when the number of pupils questioned remains at 4,

(ii) when the number of pupils questioned is increased to 8.

Give each of these answers to three decimal places.

The visitor puts the same question to each of 4 pupils chosen at random from a class of 10 pupils, 8 of whom like mathematics. Calculate the probabilities of obtaining the answer 'yes' from 0, 1, 2, 3, 4 pupils. *(JMB)*

13 Based on statistical evidence, a life insurance company uses the following table for the calculation of premiums. The table shows, per thousand of the population, the number of persons expected to survive to a particular age.

Age	0	10	20	30	40	50	60	70	80	90	100
No. of survivors	1000	980	966	949	920	874	770	539	260	40	0

Use the table to estimate the following probabilities.

(i) That a person born now will die in the next 10 years.

(ii) That a person born now will survive to age 60.

(iii) That a person born now will die between the ages of 30 and 40.

(iv) That a person aged 40 will die in the next 10 years.

The probability that a person aged 60 will die within the next 10 years is 0.3. The company has a life policy for people aged 60 such that if a person aged 60 dies within the next 10 years his family gets £5000. Survivors get nothing. Calculate the single premium that a company should charge so as to break even in the long run. (Ignore any interest on premiums and any administrative costs.)

Twenty people take out this policy. The company charges each person a single premium of £1800. Using the statistical tables provided, find the probability that the company will make a loss. (*JMB*)

14 Let X denote the number of heads obtained when a fair coin is tossed three times. Write down the mean and the variance of X.

A pack of six cards consists of three cards labelled H and the other three labelled T. Three cards are drawn at random without replacement from the pack. Let Y denote the number of H's obtained. Calculate the probabilities that Y takes the values 0, 1, 2 and 3 respectively. Show that the mean of Y is equal to the mean of X and express the variance of Y as a percentage of the variance of X. (*JMB*)

15 Define the binomial distribution, stating the conditions under which it will arise, and find its mean.

A manufacturer of glass marbles produces equal large numbers of red and blue marbles. These are thoroughly mixed together and then packed in packets of six marbles which are random samples from the mixture. Find the probability distribution of the number of red marbles in a packet purchased by a boy.

Two boys, Fred and Tom, each buy a packet of marbles. Fred prefers the red ones and Tom the blue ones, so they agree to exchange marbles as far as possible, in order that at least one of them will have six of the colour he prefers. Find the probabilities that, after the exchange,

(i) they will both have a set of six of the colour they prefer,

(ii) Fred will have three or more blue ones. (*JMB*)

16 A trial may have two outcomes, success or failure. If in n such independent trials, the probability p of a success remains constant from trial to trial, write down the probability of r successes in the n trials.

When two friends A and B play chess, the probability that A wins any game is $\frac{2}{5}$, and if A does not win the game, the probabilities then of B winning and of a draw are equal. In the course of an evening they play four games. Calculate the probabilities

(i) that A does not win a game,

(ii) that he wins more than two games.

If it is known that A has won exactly two of these four games, write down the probability distribution of the number of games that B has won.

Calculate the probability that A wins more games than B when four games are played.

(JMB)

17 For each of the books sent out by a book club, the probability of retention by a member to whom it is sent is $\frac{3}{4}$ and the probability of return to the club is $\frac{1}{4}$.

For each book returned there is a probability of $\frac{1}{3}$ that it is so damaged as to be unusable. Returned books that are not unusable are sent out to other members, the probabilities of return and of damage being the same as before. Calculate the probability that a book is returned and unusable after being sent out once. Calculate also the probability that it is sent out twice and is then returned and unusable. Show that the probability that a book is eventually returned and unusable is $\frac{1}{10}$.

For a random sample of five books write down, in fractional form, the probability distribution of the number that will eventually be returned and unusable. Calculate the probabilities that this number is

(i) more than three,

(ii) less than three.

(JMB)

18 A small factory has 15 employees of whom 3 are office staff. Each month for three months one worker is chosen at random and presented with a selection of the firm's products. Write down the probability distribution of the number of times the winner is a member of the office staff

(i) if previous winners are eligible again,

(ii) if previous winners are not eligible again.

For each method of selection calculate the mean number of times that the office staff win.

19 A battery used to operate a device has n cells, each of which, independently of the other cells, has probability p of being functional. When r of the cells are functional, the probability that the battery will operate the device is $\dfrac{r}{n}$, for $0 \leqslant r \leqslant n$.

(i) Let X denote the number of functional cells in the battery. Name the distribution of X and write down expressions for $E[X]$ and $E[X^2]$.

(ii) Obtain an expression for the joint probability that the battery will operate the device and have exactly r functional cells. Hence show that the probability that the battery will operate the device is p.

(iii) Given that the battery does not operate the device, find an expression for the probability that the battery has exactly r functional cells. Hence, or otherwise, show that when the device is operating, the expected value of the number of functional cells in the battery is
$$1 - p + np. \qquad (JMB)$$

20 When a market gardener takes n cuttings from a shrub and plants them, the number X that will root successfully is a discrete random variable with
$$P(X = r) = \frac{2r}{n(n + 1)} \qquad r = 1, 2, \ldots, n.$$

The total cost in pence to the gardener of taking and planting n cuttings is equal to $20 + 0.8n^2$. Cuttings that root successfully are sold by the gardener for 60 pence each.

a Show that $E[X] = \dfrac{(2n + 1)}{3}$.

b For $n = 20$, calculate
(i) the probability that the gardener will make a loss,
(ii) the gardener's expected profit.

c Determine the value of n which will maximise the gardener's expected profit and evaluate this maximum expected profit. $\qquad (WJEC)$

21 A loaded die is such that when it is thrown the probability of obtaining a 5 is p and the probability of obtaining a 6 is also p, where $0 < p < \dfrac{1}{2}$.

a If the die is thrown n times
(i) find, in terms of n and p, the probability that no 5 or 6 will be obtained;
(ii) find, in terms of n and p, the probability that at least one 5 or at least one 6 will be obtained;
(iii) show that the probability of obtaining at least one 5 and no 6 is
$$(1 - p)^n - (1 - 2p)^n.$$

b If the die is to be thrown until both a 5 and a 6 have been obtained, show that the mean number of throws that will have to be made is equal to $\dfrac{3}{(2p)}$.

[You may assume that for $|x| < 1$, $\displaystyle\sum_{r=1}^{\infty} rx^{r-1} = (1 - x)^{-2}$.] $\qquad (JMB)$

Chapter 8

Continuous random variables

In Chapter 6 the idea of a random variable was introduced and the properties of discrete random variables studied. In this chapter those ideas will be extended to include the properties of continuous random variables. A continuous random variable is a random variable whose range space is neither finite nor countably infinite, i.e. the elements of its range space cannot be listed.

8.1 Probability density functions

Let X be a continuous random variable with range space $R_x = \{x : a \leqslant x \leqslant b\}$. Let f be the function with domain R_x such that

$$f(x) \geqslant 0$$

$$\int_a^b f(x)\,dx = 1$$

$$\int_c^d f(x)\,dx = P(c \leqslant X \leqslant d)$$

for all c, d such that $a \leqslant c \leqslant d \leqslant b$. The function f is called the probability density function of X.

It should be noted that it follows from the above definition that

$$P(X = c) = 0$$

and

$$P(c < X < d) = P(c \leqslant X < d) = P(c < X \leqslant d) = P(c \leqslant X \leqslant d).$$

It is also important to note that the value of $f(x)$ itself does not represent a probability, it represents the density of the probability in the neighbourhood of the value of x.

The graph of a probability density function f has properties analogous to those of a relative frequency histogram. For example, the area of the region bounded by the graph of f, the ordinates $x = a$, $x = b$ and the x-axis is equal to unity, which is also the total area of the relative frequency histogram. Again, the area of the region bounded by the graph of f, the ordinates $x = c$, $x = d$ and the x-axis is equal to $P(c < X < d)$, whilst the area of the relative frequency histogram between $x = c$ and $x = d$ is equal to the relative frequency of the values of x between c and d.

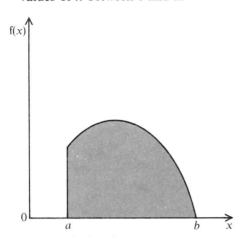

The shaded area = 1.

The shaded area represents $P(c < X < d)$.

Example 1

The continuous random variable X has the probability density function f given by

$$f(x) = kx(4 - x) \qquad 0 \leqslant x \leqslant 4,$$
$$f(x) = 0 \qquad \text{otherwise.}$$

Find   **a** the value of k **b** $P(X \leqslant 2)$ **c** $P(1 < X < 3)$.

a   Since $\displaystyle\int_a^b f(x)\,dx = 1$

$$\int_0^4 kx(4 - x)\,dx = 1$$

$$k\int_0^4 (4x - x^2)\,dx = 1$$

$$k\left[2x^2 - \frac{x^3}{3}\right]_0^4 = 1$$

$$k = \frac{3}{32}$$

130

b
$$P(X \leqslant 2) = \int_0^2 kx(4 - x)\,dx$$

$$= \frac{3}{32}\left[2x^2 - \frac{x^3}{3}\right]_0^2$$

$$= \frac{1}{2}$$

c
$$P(1 < X < 3) = \int_1^3 kx(4 - x)\,dx$$

$$= \frac{3}{32}\left[2x^2 - \frac{x^3}{3}\right]_1^3$$

$$= \frac{11}{16}$$

Exercise 8.1

1 The random variable X has the probability density function f given by
$$f(x) = k(2 - x) \qquad 0 \leqslant x \leqslant 2,$$
$$f(x) = 0 \qquad \text{otherwise.}$$
Find   **a** the value of k **b** $P(X < 1)$ **c** $P\left(X \geqslant \frac{1}{2}\right)$.

2 The random variable X has the probability density function f given by
$$f(x) = kx(4 - x) \qquad 0 \leqslant x \leqslant 3,$$
$$f(x) = 0 \qquad \text{otherwise.}$$
Find   **a** the value of k **b** $P(X \leqslant 2)$ **c** $P(1 < X < 2)$.

3 The random variable X has the probability density function f given by
$$f(x) = k \qquad 4 \leqslant x \leqslant 8,$$
$$f(x) = 0 \qquad \text{otherwise.}$$
Find   **a** the value of k **b** $P(X < 6)$ **c** $P(5 \leqslant X \leqslant 7)$.

4 The random variable X has the probability density function f given by
$$f(x) = kx^3 \qquad 0 \leqslant x \leqslant 2,$$
$$f(x) = 0 \qquad \text{otherwise.}$$
Find   **a** the value of k **b** $P(X > 1)$ **c** $P(X > 3)$.

5 Explain why the function f given by
$$f(x) = \frac{x(5 - x)}{8} \qquad 0 \leqslant x \leqslant 6,$$
$$f(x) = 0 \qquad \text{otherwise,}$$
cannot be a probability density function.

8.2 Expected value and variance

If X is a continuous random variable with range space $R_x = \{x: a \leqslant x \leqslant b\}$ and probability density function f, then the *expected value* (or *expectation*) of X is given by

$$E[X] = \int_a^b xf(x)\,dx.$$

The expected value of X is also called the *mean* of the distribution of X and is often denoted by μ.

Expected value of a function

The *expected value* (or *expectation*) of $g(X)$, where g is a function, is given by

$$E[g(X)] = \int_a^b g(x)f(x)\,dx.$$

The following results may be easily proved using the properties of integrals

$$E[b] = b$$

$$E[aX] = aE[X]$$

$$E[aX + b] = aE[X] + b$$

$$E[ag(X) + bh(X)] = aE[g(X)] + bE[h(X)].$$

The proofs are left as an exercise for the reader. These results are identical in form to the corresponding results for discrete random variables given on page 100 in Chapter 6.

Variance

The *variance* $V[X]$ of the distribution of a continuous random variable X, whose mean is μ, is given by

$$V[X] = E[(X - \mu)^2].$$

The calculation of the value of $V[X]$ may be facilitated by the use of an alternative formula derived below.

$$\begin{aligned}
V[X] &= \int_a^b (x - \mu)^2 f(x)\,dx \\
&= \int_a^b (x^2 - 2\mu x + \mu^2)f(x)\,dx \\
&= \int_a^b x^2 f(x)\,dx - 2\mu \int_a^b xf(x)\,dx + \mu^2 \int_a^b f(x)\,dx \\
&= E[X^2] - 2\mu E[X] + \mu^2 \\
&= E[X^2] - \mu^2
\end{aligned}$$

or $\qquad V[X] = E[X^2] - \{E[X]\}^2$

This is identical in form to the corresponding formula for a discrete random variable given on page 101 in Chapter 6. The standard deviation, denoted by σ or SD[X], is again defined as the positive square root of V[X].

Example 2

Find the mean and the variance of the distribution of X, whose probability density function f is given by

$$f(x) = \frac{3x(4 - x)}{32} \qquad 0 \leqslant x \leqslant 4,$$

$$f(x) = 0 \qquad \text{otherwise.}$$

$$E[X] = \int_a^b xf(x)\,dx$$

$$= \int_0^4 \frac{3}{32}x^2(4 - x)\,dx$$

$$= \frac{3}{32}\int_0^4 (4x^2 - x^3)\,dx$$

$$= \frac{3}{32}\left[\frac{4x^3}{3} - \frac{x^4}{4}\right]_0^4$$

Therefore $E[X] = 2$ or $\mu = 2$.

Since the graph of f is symmetrical about the line $x = 2$ it is possible to conclude that $\mu = 2$ without performing the integration above.

$$E[X^2] = \int_a^b x^2 f(x)\,dx$$

$$= \int_0^4 \frac{3}{32}x^3(4 - x)\,dx$$

$$= \frac{3}{32}\int_0^4 (4x^3 - x^4)\,dx$$

$$= \frac{3}{32}\left[x^4 - \frac{x^5}{5}\right]_0^4$$

$$= \frac{24}{5}$$

$$V[X] = E[X^2] - \{E[X]\}^2$$

$$= \frac{24}{5} - 2^2$$

$$= \frac{4}{5}$$

133

Example 3

Find the expected value of X whose probability density function f is given by

$$f(x) = \frac{6x^2}{11} \qquad\qquad 0 \leqslant x \leqslant 1,$$

$$f(x) = \frac{6x}{11} \qquad\qquad 1 < x \leqslant 2,$$

$$f(x) = 0 \qquad\qquad \text{otherwise.}$$

$$E[X] = \int_a^b xf(x)\,dx$$

$$= \int_0^1 x\frac{6x^2}{11}\,dx + \int_1^2 x\frac{6x}{11}\,dx$$

$$= \int_0^1 \frac{6x^3}{11}\,dx + \int_1^2 \frac{6x^2}{11}\,dx$$

$$= \left[\frac{6x^4}{44}\right]_0^1 + \left[\frac{6x^3}{33}\right]_1^2$$

$$= \frac{3}{22} + \frac{16}{11} - \frac{2}{11}$$

$$= \frac{31}{22}$$

Exercise 8.2

In each of the following, find the mean and the variance of the random variable X whose probability density function is f.

1 $f(x) = \dfrac{(2 - x)}{2}$ $0 \leqslant x \leqslant 2,$

  $f(x) = 0$ otherwise.

2 $f(x) = \dfrac{x(4 - x)}{9}$ $0 \leqslant x \leqslant 3,$

  $f(x) = 0$ otherwise.

3 $f(x) = \dfrac{1}{4}$ $4 \leqslant x \leqslant 8,$

  $f(x) = 0$ otherwise.

4 $f(x) = \dfrac{x^3}{4}$ $\qquad\qquad 0 \leqslant x \leqslant 2,$

$\quad f(x) = 0$ $\qquad\qquad$ otherwise.

5 $f(x) = \dfrac{3x^2}{7}$ $\qquad\qquad 0 \leqslant x \leqslant 1,$

$\quad f(x) = \dfrac{3}{7}$ $\qquad\qquad 1 < x \leqslant 3,$

$\quad f(x) = 0$ $\qquad\qquad$ otherwise.

8:3 Cumulative distribution function

For any random variable X, the function F, which is such that

$$F(x) = P(X \leqslant x) \qquad\qquad \text{for all } x,$$

is called the *cumulative distribution function* (or simply the *distribution function*) of X.

For a continuous random variable X with range space $R_x = \{x : a \leqslant x \leqslant b\}$ and probability density function f, the distribution function F is given by

$$F(x) = 0 \qquad\qquad x < a$$

$$F(x) = \int_a^x f(t)\,dt \qquad\qquad a \leqslant x \leqslant b$$

$$F(x) = 1 \qquad\qquad x > b.$$

[N.B. the letter t is a dummy variable.]

Since differentiation is the inverse operation to integration, it follows that

$$f(x) = F'(x)$$

where $F'(x)$ is the derivative of $F(x)$ with respect to x.

Median and percentiles

For a continuous random variable X with distribution function F the value m which is such that

$$F(m) = P(X \leqslant m) = 0.5$$

is called the *median* of the distribution of X. The ordinate $x = m$ divides the region enclosed by the graph of f, the ordinates $x = a$, $x = b$ and the x-axis into two halves.

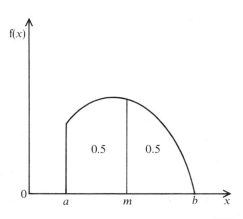

The value, x_k, which is such that

$$F(x_k) = P(X \leqslant x_k) = \frac{k}{100}$$

is called the kth *percentile* of the distribution of X. The 25th and 75th percentiles are called the *lower* and *upper quartiles* respectively. The interquartile range is used as a measure of dispersion when the median is used as a measure of location.

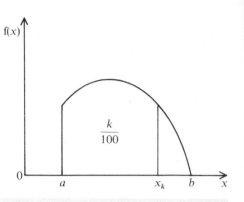

Example 4

A random variable X has the probability density function f given by

$$f(x) = \frac{6x^2}{11} \qquad 0 \leqslant x \leqslant 1,$$

$$f(x) = \frac{6x}{11} \qquad 1 < x \leqslant 2,$$

$$f(x) = 0 \qquad \text{otherwise.}$$

Find **a** the distribution function of X **b** the median **c** the 5th percentile.

a for $0 \leqslant x \leqslant 1$

$$F(x) = \int_0^x \frac{6t^2}{11} \, dt$$

$$= \left[\frac{2t^3}{11} \right]_0^x$$

$$= \frac{2x^3}{11}$$

for $1 < x \leqslant 2$

$$F(x) = \int_0^1 \frac{6t^2}{11} \, dt + \int_1^x \frac{6t}{11} \, dt$$

$$= \left[\frac{2t^3}{11} \right]_0^1 + \left[\frac{3t^2}{11} \right]_1^x$$

$$= \frac{2}{11} + \frac{3x^2}{11} - \frac{3}{11}$$

$$= \frac{3x^2 - 1}{11}$$

The distribution function

$$F(x) = 0 \qquad\qquad x < 0$$

$$F(x) = \frac{2x^3}{11} \qquad\qquad 0 \leqslant x \leqslant 1$$

$$F(x) = \frac{(3x^2 - 1)}{11} \qquad\qquad 1 < x \leqslant 2$$

$$F(x) = 1 \qquad\qquad x > 2.$$

b The median m is given by $F(m) = 0.5$.

Since $F(1) = \dfrac{2}{11} < 0.5$ it follows that m is greater than 1.

Therefore
$$\dfrac{(3m^2 - 1)}{11} = 0.5$$

$$3m^2 - 1 = 5.5$$

$$m^2 = \dfrac{6.5}{3}$$

$$m = 1.472 \text{ (3 d.p.)}$$

c The 5th percentile x_5 is given by $F(x_5) = 0.05$.

Since $F(1) = \dfrac{2}{11} > 0.05$ it follows that x_5 is less than 1.

Therefore
$$\dfrac{2x_5{}^3}{11} = 0.05$$

$$x_5{}^3 = \dfrac{0.55}{2}$$

$$x_5 = 0.650 \text{ (3 d.p.)}$$

Exercise 8.3

In each of the following, given that f is the probability density function of a continuous random variable X, find the distribution function and the median. In questions **1** to **4** also find the interquartile range.

1 $f(x) = \dfrac{x}{2}$ $0 \leqslant x \leqslant 2$,

   $f(x) = 0$ otherwise.

2 $f(x) = 3(1 - x)^2$ $0 \leqslant x \leqslant 1$,
   $f(x) = 0$ otherwise.

3 $f(x) = \dfrac{1}{x^2}$ $1 \leqslant x \leqslant 3$,

   $f(x) = \dfrac{1}{9}$ $3 < x \leqslant 6$,

   $f(x) = 0$ otherwise.

4 $f(x) = \dfrac{1}{2x^2}$ $1 \leqslant x \leqslant 3$,

   $f(x) = \dfrac{1}{18}$ $3 < x \leqslant 15$,

   $f(x) = 0$ otherwise.

5 $f(x) = 6x(1 - x)$ $0 \leqslant x \leqslant 1$,
   $f(x) = 0$ otherwise.

8.4 Some special distributions

The most important type of continuous distribution is the normal distribution; this will be studied in detail in Chapter 9.

Continuous uniform distribution

The continuous random variable X which has the probability density function f given by

$$f(x) = \frac{1}{(b-a)} \qquad a \leqslant x \leqslant b$$

$$f(x) = 0 \qquad \text{otherwise}$$

is said to be uniformly (or rectangularly) distributed between a and b. This is often abbreviated as $X \sim U(a, b)$.

The mean of the distribution is given by

$$E[X] = \int_a^b x \cdot \frac{1}{(b-a)} \, dx$$

$$= \frac{1}{(b-a)} \left[\frac{x^2}{2} \right]_a^b$$

$$= \frac{1}{2(b-a)} (b^2 - a^2)$$

$$= \frac{(b-a)(b+a)}{2(b-a)}$$

$$= \frac{(b+a)}{2}$$

This result could have been obtained more easily by symmetry.

$$E[X^2] = \int_a^b x^2 \cdot \frac{1}{(b-a)} \, dx$$

$$= \frac{1}{(b-a)} \left[\frac{x^3}{3} \right]_a^b$$

$$= \frac{1}{3(b-a)} (b^3 - a^3)$$

$$= \frac{(b-a)(b^2 + ba + a^2)}{3(b-a)}$$

$$= \frac{1}{3}(b^2 + ba + a^2)$$

$$V[X] = E[X^2] - \{E[X]\}^2$$

$$= \frac{1}{3}(b^2 + ba + a^2) - \frac{(b + a)^2}{4}$$

$$= \frac{1}{12}(b - a)^2$$

The mean and variance of $U(a, b)$ are $\dfrac{(b + a)}{2}$ and $\dfrac{(b - a)^2}{12}$ respectively.

Example 5

A straight piece of wire AB is 20 cm long and M is its mid-point. A point P on AM is chosen at random and the wire is bent into the form of a rectangle with AP as one side. If X cm denotes the length of AP and Y cm denotes the area of the rectangle, find

a $E[X]$ **b** $E[X^2]$ **c** $E[Y]$ and   **d** $P(Y > 16)$.

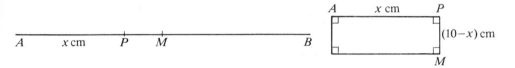

a Since P is chosen at random $X \sim U(0, 10)$

$$E[X] = \frac{(0 + 10)}{2} = 5.$$

b Since $V[X] = E[X^2] - \{E[X]\}^2$ $E[X^2] = V[X] + \{E[X]\}^2$

$$= \frac{(10 - 0)^2}{12} + 5^2$$

$$= \frac{100}{3}$$

c $Y = X(10 - X)$ $E[Y] = E[X(10 - X)]$

$$= E[10X - X^2]$$

$$= 10E[X] - E[X^2]$$

$$= 50 - \frac{100}{3}$$

$$= \frac{50}{3}$$

d $P(Y > 16) = P(10X - X^2 > 16)$

$$= P(0 > X^2 - 10X + 16)$$

$$= P(0 > (X - 2)(X - 8))$$

$$= P(2 < X < 8)$$

139

The graph of the probability density function f of X is shown below.

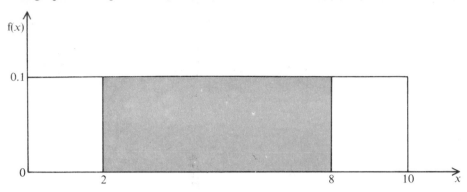

P$(2 < X < 8)$ is equal to the shaded area, therefore P$(Y > 16) = 0.6$.

Example 6

A random variable A is uniformly distributed between 3 and 9, find its distribution function.

Find the probability that the quadratic equation in x
$$x^2 - 2Ax - (3A + 10) = 0$$
has real roots.

The graph of the probability density function f of A is shown below. Since the length of the base of the rectangle is 6 and its area is 1, its height is $\frac{1}{6}$.

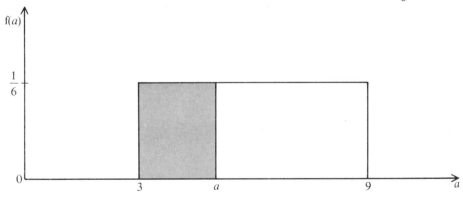

For $3 < a < 9$, P$(A \leqslant a)$ is equal to the shaded area which is $\dfrac{(a - 3)}{6}$. Thus the distribution function F of A is given by

$$F(a) = 0 \qquad\qquad\qquad a < 3$$

$$F(a) = \frac{(a - 3)}{6} \qquad\qquad 3 \leqslant a \leqslant 9$$

$$F(a) = 1 \qquad\qquad\qquad a > 9$$

P (equation has real roots) $= P(\text{“}b^2 - 4ac > 0\text{”})$

$$= P(4A^2 - 4(3A + 10) > 0)$$

$$= P(4(A^2 - 3A - 10) > 0)$$

$$= P(4(A - 5)(A + 2) > 0)$$

since A is positive $\qquad = P(A > 5)$

$$= 1 - F(5)$$

$$= 1 - \frac{(5 - 3)}{6}$$

$$= \frac{2}{3}$$

Exponential distribution

The continuous random variable X with probability density function f given by

$$f(x) = \lambda e^{-\lambda x} \qquad x > 0,$$

$$f(x) = 0 \qquad\qquad \text{otherwise,}$$

where $\lambda > 0$, is said to have the *exponential distribution* with parameter λ.

The mean of this distribution is given by

$$E[X] = \int_0^\infty x\lambda\, e^{-\lambda x}\, dx$$

integrating by parts $\qquad E[X] = \left[-xe^{-\lambda x} \right]_0^\infty - \int_0^\infty -e^{-\lambda x}\, dx$

$$E[X] = \left[-xe^{-\lambda x} \right]_0^\infty - \left[\frac{1}{\lambda}e^{-\lambda x} \right]_0^\infty$$

since $\lambda > 0$, $e^{-\lambda x} \to 0$ and $xe^{-\lambda x} \to 0$ as $x \to \infty$

therefore $\qquad\qquad E[X] = \frac{1}{\lambda}.$

Also $\qquad\qquad E[X^2] = \int_0^\infty x^2\lambda e^{-\lambda x}\, dx$

$$E[X^2] = \left[-x^2 e^{-\lambda x} \right]_0^\infty - \int_0^\infty -2xe^{-\lambda x}\, dx$$

$$E[X^2] = \left[-x^2 e^{-\lambda x} \right]_0^\infty + \frac{2}{\lambda}\int_0^\infty x\lambda e^{-\lambda x}\, dx$$

since $\lambda > 0$, $x^2 e^{-\lambda x} \to 0$ as $x \to \infty$

$$E[X^2] = \frac{2}{\lambda} E[X]$$

substituting for $E[X]$ $\qquad E[X^2] = \frac{2}{\lambda^2}$

$V[X] = E[X^2] - \{E[X]\}^2 \qquad V[X] = \frac{2}{\lambda^2} - \frac{1}{\lambda^2}$

$$V[X] = \frac{1}{\lambda^2},$$

Thus the mean and the variance of the exponential distribution are $\frac{1}{\lambda}$ and $\frac{1}{\lambda^2}$ respectively.

Two examples of the practical use of the exponential distribution are when it is used as a model for:
(i) the distribution of the operational lifetimes of certain types of electronic components
(ii) the distribution of the times between successive calls at a telephone switchboard.

Example 7

The operational lifetime X of an electronic component has an exponential distribution with parameter λ, find its distribution function and show that

$$P(X > a + b \,|\, X > a) = P(X > b).$$

$$F(x) = \int_0^x \lambda e^{-\lambda t}\, dt$$

$$F(x) = \left[-e^{-\lambda t} \right]_0^x$$

Distribution function $\qquad F(x) = 1 - e^{-\lambda x} \qquad\qquad x > 0$

$$F(x) = 0 \qquad\qquad x \leqslant 0$$

$$P(X > a + b \,|\, X > a) = \frac{P(X > a + b)}{P(X > a)}$$

$$= \frac{1 - F(a + b)}{1 - F(a)}$$

$$= \frac{1 - (1 - e^{-\lambda(a + b)})}{1 - (1 - e^{-\lambda a})}$$

$$= \frac{e^{-\lambda(a + b)}}{e^{-\lambda a}}$$

$$= e^{-\lambda b}$$

$P(X > b) = 1 - F(b)$

$\qquad = 1 - (1 - e^{-\lambda b})$

$\qquad = e^{-\lambda b}$

Hence $\qquad P(X > a + b \,|\, X > a) = P(X > b)$.

Thus when the component is operational the probability that it will operate for a further time b is the same as the probability that it will operate for a time b from new. This is an example of the so called 'lack of memory' of the exponential distribution.

The next two examples illustrate further techniques in questions concerned with continuous random variables.

Example 8

A random variable T has the probability density function given by

$$f(t) = \frac{t}{50}, \qquad\qquad 0 \leqslant t \leqslant 10,$$

$$f(t) = 0, \qquad\qquad \text{otherwise.}$$

Find the distribution function of T.

A particle P moves along a straight line such that its velocity v at a time t is given by $v = 3t^2$. The particle is observed at time t, where t denotes the value of a random variable whose probability density function is the function f defined above. Calculate the probabilities that at the time of observation

a the speed of P is less than 147

b the distance travelled by P is at least 512.

For $0 \leqslant t \leqslant 10$, $\qquad\qquad F(t) = \int_0^t \frac{y}{50}\,dy$

$$= \left[\frac{y^2}{100}\right]_0^t$$

$$= \frac{t^2}{100}$$

Distribution function $\qquad F(t) = 0 \qquad\qquad\qquad t < 0$

$$F(t) = \frac{t^2}{100} \qquad\qquad 0 \leqslant t \leqslant 10$$

$$F(t) = 1 \qquad\qquad\qquad t > 10$$

a $P(V < 147) = P(3T^2 < 147)$

$\qquad\qquad = P(T < 7)$

$\qquad\qquad = F(7)$

$\qquad\qquad = 0.49$

b $v = 3t^2$

integrating w.r.t. t $x = t^3 + k$

when $t = 0$, $x = 0$ therefore $k = 0$

$$x = t^3$$

$$P(X > 512) = P(T^3 > 512)$$
$$= P(T > 8)$$
$$= 1 - F(8)$$
$$= 0.36$$

Example 9

A continuous random variable X has the probability density function f given by

$$f(x) = \frac{3x(4 - x)}{32}, \qquad 0 \leqslant x \leqslant 4,$$

$$f(x) = 0, \qquad\qquad \text{otherwise.}$$

Find, to three decimal places, the probability that at least one of five independently chosen values of X will exceed 3.

For one value of X, $P(X > 3) = \displaystyle\int_3^4 \frac{3}{32}x(4 - x)\,dx$

$$= \frac{3}{32}\left[2x^2 - \frac{x^3}{3} \right]_3^4$$

$$= \frac{5}{32}$$

Let Y be the number of values of X out of 5 which exceed 3.

$$Y \sim B\left(5, \frac{5}{32} \right)$$

$$P(Y \geqslant 1) = 1 - P(Y = 0)$$

$$= 1 - \left(\frac{27}{32} \right)^5$$

$$= 0.572 \ (3 \text{ d.p.})$$

Miscellaneous Exercise ⟨ 8 ⟩

1 A continuous random variable X has probability density function f given by

$$f(x) = 1 - \frac{x}{2}, \qquad\qquad 0 \leqslant x \leqslant 2,$$

$$f(x) = 0, \qquad\qquad \text{otherwise.}$$

Find the exact values of the mean and the median of X.   (*JMB*)

2 The probability density function f of a continuous random variable X is given by

$$f(x) = kx, \qquad 6 \leqslant x \leqslant 10,$$

$$f(x) = 0, \qquad \text{otherwise.}$$

Find the mean and median values of X.

3 The probability density function of x is given by

$$f(x) = k(ax - x^2), \qquad 0 \leqslant x \leqslant 2,$$

$$f(x) = 0, \qquad x < 0, \quad x > 2,$$

where k and a are positive constants.

Show that $a \geqslant 2$ and that $k = \dfrac{3}{6a - 8}$.

Given that the mean value of x is 1, calculate the values of a and k.

For these values of a and k sketch the graph of the probability density function and find the variance of x. (*JMB*)

4 A continuous random variable X has the probability density function f defined by

$$f(x) = \frac{cx}{3}, \qquad 0 \leqslant x < 3,$$

$$f(x) = c, \qquad 3 \leqslant x \leqslant 4,$$

$$f(x) = 0, \qquad \text{otherwise,}$$

where c is a positive constant.

Find (i) the value of c,

(ii) the mean of X,

(iii) the value, a, for there to be a probability of 0.85 that a randomly chosen value of X will exceed a. (*JMB*)

5 The times, t, between consecutive breakdowns of a machine are independent and have probability density function $Ae^{-\lambda t}$, $t > 0$, where λ is a positive constant. Find A in terms of λ.

Show that the mean of this distribution is $\dfrac{1}{\lambda}$. Demonstrate that the probability that the time between consecutive breakdowns is less than the mean does not depend on the value of the mean. Calculate this probability (to 3 decimal places).

Sketch the distribution for the values $\lambda = 0.5$, 1 and 2, using the same coordinate axes for all three cases, and mark the position of the mean in each case. (*JMB*)

6 The continuous random variable X has the probability density function f given by

$$f(x) = kx, \qquad\qquad 5 < x < 10,$$
$$f(x) = 0, \qquad\qquad \text{otherwise.}$$

(i) Find the value of k.

(ii) Find the expected value of X.

(iii) Find the probability that $X > 8$.

The annual income from money invested in a Unit Trust Fund is X per cent of the amount invested, where X has the above distribution. Suppose you have a sum of money to invest and that you are prepared to leave your money invested over a period of several years. State, with your reasons, whether you would invest in the Unit Trust Fund or in a Money Bond offering a guaranteed annual income of 8 per cent on money invested.

(*JMB*)

7 A continuous random variable has probability density function f defined by

$$f(x) = kx(4 - x) \qquad\qquad 0 \leqslant x \leqslant 4,$$
$$f(x) = 0 \qquad\qquad\qquad \text{otherwise.}$$

Evaluate k and the mean of the distribution.

A particle moves along a straight line in such a way that during the first four seconds of its motion its velocity at time t seconds is $v\,\text{m/s}$, where

$$v = 2(t + 1).$$

The particle is observed at time t seconds where t denotes a value of a random variable whose probability density function is the function f defined above. Calculate the probabilities that at the time of observation

(i) the velocity of the particle is less than $4\,\text{m/s}$

(ii) the particle is at least $3\,\text{m}$ from its starting point. (*JMB*)

8 The probability density function of x is

$$f(x) = kx(1 - ax^2), \qquad\qquad 0 \leqslant x \leqslant 1,$$
$$f(x) = 0, \qquad\qquad\qquad x < 0, \quad x > 1,$$

where k and a are positive constants. Show that $a \leqslant 1$ and that $k = \dfrac{4}{2 - a}$.

Sketch the probability density function for the case when $a = 1$ and in this case find the mean, variance and 75th percentile of this distribution.

(*JMB*)

9 The number of cubic centimetres of a certain ingredient contained in a litre of a raw material is a continuous random variable X whose probability density function is given by

$$f(x) = cx, \qquad\qquad 0 < x \leqslant 5,$$
$$f(x) = c(10 - x), \qquad 5 < x \leqslant 10,$$
$$f(x) = 0, \qquad\qquad \text{otherwise.}$$

(i) Find the value of the constant c.

(ii) Determine the mean and the variance of X.

(iii) The cost of extracting the ingredient from 1 litre of the raw material is £1 if $0 < x \leqslant 3$, £2 if $3 < x \leqslant 6$, and £3 if $6 < x \leqslant 10$. Calculate the expected cost of extracting the ingredient from 1 litre of the raw material.

(JMB)

10 A continuous random variable X has a distribution whose probability density function f is given by

$$f(x) = kx, \qquad 0 \leqslant x \leqslant 10,$$

$$f(x) = 0, \qquad \text{otherwise.}$$

(i) Calculate the value of k.

(ii) Calculate the mean and variance of the distribution.

(iii) Let x denote a randomly observed value from the distribution. Find an expression for the probability that x will be in the range from $n + 0.5$ up to but excluding $n + 0.6$ where n is an integer from 0 to 9 inclusive. Hence calculate the probability that the first digit after the decimal point in x is equal to 5.

(JMB)

11 Independently for each shot aimed by Alec at the centre of a board, the point at which the shot strikes the board is at a distance X cm from the centre of the board, where X is a continuous random variable having probability density function f defined by

$$f(x) = \frac{2(5 - x)}{25}, \qquad 0 < x < 5,$$

$$f(x) = 0, \qquad \text{otherwise.}$$

Calculate the mean and the variance of X.

A shot striking the board within 0.5 cm of the centre is a success.

(i) Show that the probability that Alec's first attempt will be a success is 0.19.

(ii) Calculate, correct to three significant figures, the probability that Alec will have at least one success in three attempts.

(iii) Calculate, correct to three significant figures, the probability that Alec's second success will be on his fourth attempt.

(iv) Denoting by Y the number of successes obtained by Alec in ten attempts calculate the mean and the variance of Y.

(JMB)

12 A greengrocer sells apples singly at 10p each and by weight at 75p per kilogram. On any one day the number, R, of apples sold singly is a discrete random variable having the binomial distribution

$$P(R = r) = \binom{10}{r}(0.8)^r(0.2)^{10-r}, \qquad r = 0, 1, 2, \ldots, 10,$$

and the weight, X kg, of apples sold by weight is a continuous random variable whose probability density function is given by

$$f(x) = 0.08(x - 5), \qquad 5 \leqslant x \leqslant 10.$$

Using tables, or otherwise, find, to two decimal places, the probabilities that in a day the number of apples sold singly will be
(i) 8 or more, (ii) 8 or less, (iii) exactly 8.

Find the probabilities that in a day the weight of apples sold by weight will be
(iv) 8 kg or more, (v) less than 6 kg.

Calculate the expected daily receipts from the sale of apples. (*JMB*)

13 Petrol is delivered to a garage every Monday morning. The weekly demand for petrol at this garage, in thousands of gallons, is a continuous random variable X distributed with a probability density function of the form

$$f(x) = kx(c - 2x), \qquad \frac{1}{2} \leqslant x \leqslant \frac{3}{2},$$

$$f(x) = 0, \qquad \text{otherwise.}$$

(i) Given that the mean weekly demand is 900 gallons, determine the values of k and c.
(ii) Calculate the mean number of gallons *sold* per week at this garage if its supply tanks are filled to their total capacity of 1000 gallons every Monday morning. (*WJEC*)

14 The operating lifetimes, X hours, of a certain brand of battery are distributed with probability density function

$$f(x) = \frac{(x - 30)(70 - x)}{9000}, \qquad 40 \leqslant x \leqslant 70,$$

$$f(x) = 0, \qquad \text{otherwise.}$$

A battery whose operating lifetime is less than 50 hours is classified as being of poor quality, one whose lifetime is from 50 hours to 60 hours as being of average quality and one whose lifetime exceeds 60 hours as being of good quality. Show that the proportions of poor, average and good quality batteries are in the ratio $11:11:5$.

A simple electronics test has been devised in an attempt to predict the quality of a new battery. A battery subjected to this test may give either a positive or a negative response and it has been observed from applications of the test to new batteries that the probabilities of a positive response from a poor, an average and a good quality battery are respectively $\frac{1}{4}, \frac{1}{2}$ and $\frac{3}{4}$.

On the basis of these results, find the probability that a new battery which gives a positive response when tested is of good quality. (*WJEC*)

15 A quality characteristic X of a manufactured item is a continuous random variable having probability density function

$$f(x) = 2\lambda^{-2}x, \qquad 0 < x < \lambda,$$

$$f(x) = 0, \qquad \text{otherwise,}$$

where λ is a positive constant whose value may be controlled by the manufacturer.
(i) Find the mean and the variance of X in terms of λ.

(ii) Every manufactured item is inspected before being dispatched for sale. Any item for which X is 8 or more is passed for selling and any item for which X is less than 8 is scrapped. The manufacturer makes a profit of $£(27 - \lambda)$ on every item passed for selling, and suffers a loss of $£(\lambda + 5)$ on every item that is scrapped. Find the value of λ which the manufacturer should aim for in order to maximise his expected profit per item, and calculate his maximum expected profit per item.

$(WJEC)$

16 In the quadratic equation
$$x^2 - 2ax + a + 2 = 0$$
the value of a is randomly chosen from the interval $(-5, 5)$. Calculate the probability that the equation has real roots.

If, instead, a is randomly chosen from the interval $(2, 5)$, calculate the probability that the smaller root is less than unity.

(JMB)

17 A continuous random variable X has the probability density function f given by
$$f(x) = kx(4 - x), \qquad 0 \le x \le 3,$$
$$f(x) = 0, \qquad \text{otherwise.}$$
Show that $k = \dfrac{1}{9}$ and determine

(i) the mean and variance of X,

(ii) the distribution function of X,

(iii) the probability, to three decimal places, that at least one of five independently observed values of X will exceed 2.

(JMB)

18 A person frequently makes telephone calls to destinations for which each call is charged at the rate of 15p per minute or part of a minute. The cost of such a call is Xp, and its duration, T minutes, has the exponential probability density function
$$f(t) = \alpha e^{-\alpha t}, \qquad t \ge 0,$$
$$f(t) = 0, \qquad t < 0.$$
Show that $P(X = 15r) = e^{-r\alpha}(e^{\alpha} - 1), \qquad r = 1, 2, 3, \ldots,$

and that the mean cost per call in pence is $\dfrac{15}{(1 - e^{-\alpha})}$.

$$\left[\text{You may assume that } \sum_{r=1}^{\infty} rx^r = \frac{x}{(1 - x)^2}, \qquad |x| < 1. \right]$$

When a caller telephones a particular company, there is a probability of $\dfrac{1}{2}$ that he will be asked to hold the line. When he is asked to hold the line and he decides to do so, the total time taken for the call has the above exponential distribution with $\alpha = \frac{1}{6}$; when he is not asked to hold the line the total time for the call has the above exponential distribution with $\alpha = \frac{1}{2}$. Calculate the expected cost if he rings the company and is asked to hold the line and he (i) holds and completes the call,   (ii) rings off and then rings later, completing the second call whether he is asked to hold or not. Assume that a wasted call cost 15p and take $e^{-\frac{1}{2}} = 0.6065$, $e^{-\frac{1}{6}} = 0.8465$.

(JMB)

19 The continuous random variable X has probability density function f given by

$$f(x) = a \qquad\qquad\qquad 0 < x < 1$$
$$f(x) = b(4 - x) \qquad\qquad 1 \leqslant x \leqslant 4$$
$$f(x) = 0 \qquad\qquad\qquad \text{otherwise,}$$

where a and b are constants. Given that the mean of the distribution is $\frac{7}{5}$, show that a and b must have the values $\frac{2}{5}$ and $\frac{2}{15}$, respectively.

Sketch the graph of f(x), and determine the median value of X, giving your answer correct to two decimal places.

(JMB)

20 A tank with vertical faces has a horizontal rectangular base with sides X m and $(10 - X)$ m, where X is a continuous random variable having a uniform distribution between 6 and 9. Water is poured into the tank. Given that the volume of water in the tank is $100\,\text{m}^3$, calculate, to three decimal places, the probability that the depth of water is greater than 5 m.

Also calculate, to the nearest m^3, the least volume of water that the tank must contain for the probability to be at least 0.9 that the depth of water exceeds 5 m.

(JMB)

Chapter 9

Normal distribution

The most important type of continuous distribution in statistical theory and practice is the normal distribution. The distributions of many random variables are either normal or approximately normal; the measurements of many characteristics of living organisms, the errors of observations in certain experiments and the means of large samples are examples of such random variables. The central limit theorem, which will be discussed in Book 2, helps to explain why the normal distribution occurs so frequently in statistical work.

In the work that follows it is necessary to assume that

$$\int_{-\infty}^{\infty} \exp\left(\frac{-z^2}{2}\right) dz = \sqrt{(2\pi)} \tag{1}$$

the proof of which requires mathematics beyond A level standard.

[N.B. The notation $\exp(x)$ is used to represent e^x.]

9.1 The standard normal distribution

The random variable Z with probability density function ϕ given by

$$\phi(z) = \frac{1}{\sqrt{(2\pi)}} \exp\left(\frac{-z^2}{2}\right), \qquad -\infty < z < \infty$$

is said to have the *standard normal distribution*.

The graph of ϕ:

1 is bell shaped and symmetrical about the line $z = 0$;

2 is asymptotic to the z-axis, i.e. $\phi(z) \to 0$ as $z \to \pm\infty$;

3 has two points of inflection where $z = \pm 1$;

4 is such that the area of the region bounded by the curve and the z-axis is 1.

The result **4** follows directly from **1** above and is a necessary condition for ϕ to be a probability density function.

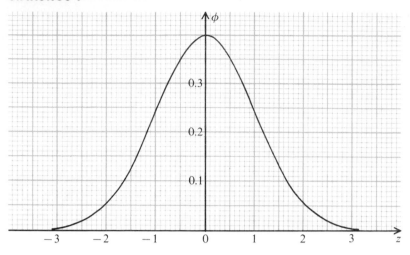

Mean and variance

$$E[Z] = \frac{1}{\sqrt{(2\pi))}} \int_{-\infty}^{\infty} z.\exp\left(\frac{-z^2}{2}\right) dz$$

Since
$$\frac{d}{dz}\left(\exp\left(\frac{-z^2}{2}\right)\right) = -z.\exp\left(\frac{-z^2}{2}\right),$$

it follows that the indefinite integral of $z.\exp\left(\frac{-z^2}{2}\right)$ is $-\exp\left(\frac{-z^2}{2}\right)$. (2)

$$E[Z] = \frac{1}{\sqrt{(2\pi)}}\left[-\exp\left(\frac{-z^2}{2}\right)\right]_{-\infty}^{\infty}$$

$$= 0$$

This result could have been established by symmetry.

$$E[Z^2] = \frac{1}{\sqrt{(2\pi)}} \int_{-\infty}^{\infty} z^2.\exp\left(\frac{-z^2}{2}\right) dz$$

Integrating by parts, taking z as one part and $z.\exp\left(\frac{-z^2}{2}\right)$ as the other, and also using result (2) above.

$$E[Z^2] = \frac{1}{\sqrt{(2\pi)}}\left(\left[-z.\exp\left(\frac{-z^2}{2}\right)\right]_{-\infty}^{\infty} - \int_{-\infty}^{\infty} -\exp\left(\frac{-z^2}{2}\right) dz\right)$$

$$= \frac{1}{\sqrt{(2\pi)}}\left[-z.\exp\left(\frac{-z^2}{2}\right)\right]_{-\infty}^{\infty} + \frac{1}{\sqrt{(2\pi)}} \int_{-\infty}^{\infty} \exp\left(\frac{-z^2}{2}\right) dz$$

The first term on the right-hand side is zero since $z.\exp\left(\frac{-z^2}{2}\right) \to 0$ as $z \to \pm\infty$, and from result (1) the second term is unity.

152

$$E[Z^2] = 1$$

Using $\quad V[Z] = E[Z^2] - \{E[Z]\}^2$

$$V[Z] = 1.$$

Thus the standard normal variable Z has mean 0 and variance 1. This is usually abbreviated to

$$Z \sim N(0, 1).$$

Distribution function of the standard normal variable

The distribution function Φ of the standard normal variable Z is given by

$$\Phi(z) = P(Z \leqslant z) = \int_{-\infty}^{z} \phi(t)\,dt$$

$$\Phi(z) = \int_{-\infty}^{z} \frac{1}{\sqrt{(2\pi)}} \exp\left(\frac{-t^2}{2}\right) dt.$$

This integral cannot be evaluated explicitly (except for $z = 0$), but it can be evaluated by numerical methods for particular values of z. The results of such evaluations have been tabulated in a number of different forms, one of which is given in **Table 3** on page 217.

Use of the table of values of the standard normal distribution function

Table 3, page 217, lists values of $\Phi(z)$ to four decimal places, for values of z from 0.00 to 3.49. Negative values of z are not listed because, for these values of z, the values of $\Phi(z)$ can be calculated from the listed values using the symmetry of the graph of ϕ and the fact that the area under the graph of ϕ is 1.

The following diagrams illustrate how various probabilities associated with values of z may be obtained using the values of $\Phi(z)$ listed in **Table 3**.

In each of the diagrams it is assumed that $0 < a < b$.

By symmetry,

(i) $\quad P(Z < a) = \Phi(a)$

(ii) $\quad P(Z > -a) = \Phi(a).$

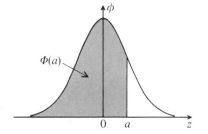

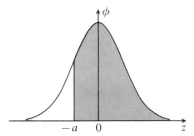

Since the total area under the graph is 1 and the unshaded area is $\Phi(a)$,

(iii) $P(Z > a) = 1 - \Phi(a)$.

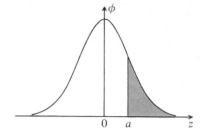

By symmetry,

(iv) $P(Z < -a) = 1 - \Phi(a)$.

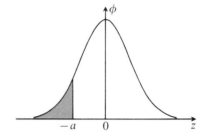

Areas to the left of $z = b$ and $z = a$ are $\Phi(b)$ and $\Phi(a)$ respectively

(v) $P(a < Z < b) = \Phi(b) - \Phi(a)$.

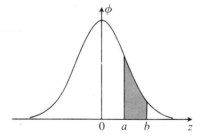

By symmetry,

(vi) $P(-b < Z < -a) = \Phi(b) - \Phi(a)$.

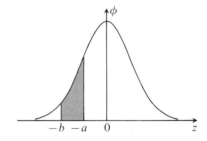

Area to the left of $z = a$ is $\Phi(a)$ and area to the left of $z = -b$ is $1 - \Phi(b)$,

(vii) $P(-b < Z < a) = \Phi(a) - (1 - \Phi(b))$
$\phantom{(vii) P(-b < Z < a)} = \Phi(a) + \Phi(b) - 1$.

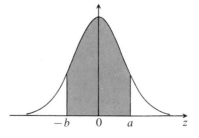

There is no need to commit all these results to memory, but it is essential to learn the following two facts:

1 The graph of ϕ is symmetrical about $z = 0$.

2 The area under the graph of ϕ is 1.

The result

$$\Phi(-a) = 1 - \Phi(a),$$

which follows from diagram (iv), is used quite often in the examples which follow.

Example 1

Given that $Z \sim N(0, 1)$, use the following extract from **Table 3**, page 217, to find

a $P(Z < 1)$ **b** $P(Z < 1.14)$ **c** $P(Z < 1.075)$

d $P(Z > 1.02)$ **e** $P(Z > -1.16)$ **f** $P(Z < -1.09)$

g $P(1.01 < Z < 1.13)$ **h** $P(-1.01 < Z < 1.13)$

i the value of a, given that $P(Z < a) = 0.87$

j the value of b, given that $P(Z < b) = 0.12$.

z	.00	0.01	.02	.03	.04	.05	.06	.07	.08	.09
1.0	.8413	.8438	.8461	.8485	.8508	.8531	.8554	.8577	.8599	.8621
1.1	.8643	.8665	.8686	.8708	.8729	.8749	.8770	.8790	.8810	.8830

It is advisable to draw a rough diagram in each case.

a $P(Z < 1) = \Phi(1.00) = 0.8413$

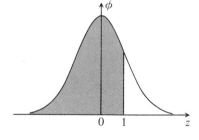

b $P(Z < 1.14) = \Phi(1.14) = 0.8729$

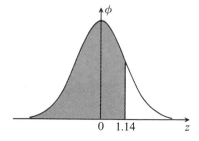

c $P(Z < 1.075) = \Phi(1.075),$

this is not tabulated but an approximate value may be found using linear interpolation.

$$P(Z < 1.075) = \Phi(1.075)$$
$$= \Phi(1.07) + 0.5[\Phi(1.08) - \Phi(1.07)]$$
$$= 0.8577 + 0.5[0.8599 - 0.8577]$$
$$= 0.8588$$

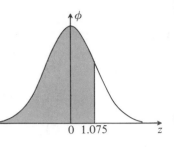

d $P(Z > 1.02) = 1 - \Phi(1.02)$
$$= 1 - 0.8461$$
$$= 0.1539$$

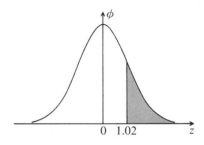

e $P(Z > -1.16) = \Phi(1.16)$
$$= 0.8770$$

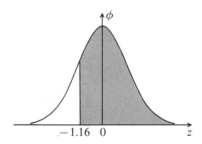

f $P(Z < -1.09) = 1 - \Phi(1.09)$
$$= 1 - 0.8621$$
$$= 0.1379$$

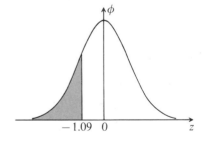

g $P(1.01 < Z < 1.13) = \Phi(1.13) - \Phi(1.01)$
$$= 0.8708 - 0.8438$$
$$= 0.0270$$

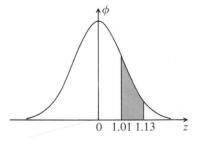

h $P(-1.01 < Z < 1.13) = \Phi(1.13) - \Phi(-1.01)$
$$= \Phi(1.13) - \{1 - \Phi(1.01)\}$$
$$= \Phi(1.13) + \Phi(1.01) - 1$$
$$= 0.8708 + 0.8438 - 1$$
$$= 0.7146$$

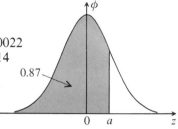

i $P(Z < a) = 0.87$
$\Phi(a) = 0.87$
$\Phi(1.12) = 0.8686$ and $\Phi(1.13) = 0.8708$,
   a is between 1.12 and 1.13
the difference between 0.8686 and 0.8708 is 0.0022
the difference between 0.8686 and 0.87 is 0.0014
using linear interpolation

$$a \simeq 1.12 + \frac{0.0014}{0.0022}[1.13 - 1.12]$$

$$a \simeq 1.126 \quad (3 \text{ decimal places})$$

j $P(Z < b) = 0.12$
$\Phi(b) = 0.12$
the value of b is clearly negative
$\Phi(-b) = 1 - \Phi(b)$
$\Phi(-b) = 1 - 0.12$
$\Phi(-b) = 0.88$
since $\Phi(1.17) = 0.8790$ and $\Phi(1.18) = 0.8810$
$-b = 1.175$
$b = -1.175$

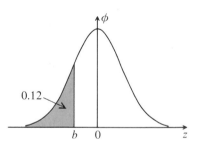

Exercise 9.1

Given that $Z \sim N(0, 1)$ use tables to evaluate the following.

1 a $P(Z < 0.8)$ **b** $P(Z < 1.37)$
   c $P(Z < 0.475)$ **d** $P(Z < 1.903)$

2 a $P(Z > 1.4)$ **b** $P(Z > 1.68)$
   c $P(Z > 1.925)$ **d** $P(Z > 0.677)$

3 a $P(Z < -0.5)$ **b** $P(Z < -1.13)$
   c $P(Z < -1.565)$ **d** $P(Z < -2.238)$

4 a $P(Z > -2.4)$ **b** $P(Z > -1.84)$
   c $P(Z > -0.355)$ **d** $P(Z > -1.022)$

5 a $P(1.61 < Z < 1.74)$ **b** $P(0.52 < Z < 0.74)$

  c $P(-1.3 < Z < 1.8)$ **d** $P(-1.96 < Z < 1.96)$

Find the values of a, b, c, d in the following.

6 a $P(Z < a) = 0.9484$ **b** $P(Z > b) = 0.0600$

  c $P(Z > c) = 0.9744$ **d** $P(Z < d) = 0.1500$

7 a $P(Z < a) = 0.8749$ **b** $P(Z > b) = 0.2296$

  c $P(Z > c) = 0.5910$ **d** $P(Z < d) = 0.1200$

8 a $P(Z < a) = 0.95$ **b** $P(Z > b) = 0.1$

  c $P(Z > c) = 0.975$ **d** $P(Z < d) = 0.01$

9.2 The general normal distribution

Consider the random variable X which is related to the standard normal variable Z by the equation

$$X = \sigma Z + \mu$$

where σ, μ are constants and $\sigma > 0$.

$$E[X] = E[\sigma Z + \mu]$$
$$= \sigma E[Z] + \mu$$
$$= \mu.$$

Therefore the mean of X is μ.

$$V[X] = V[\sigma Z + \mu]$$
$$= \sigma^2 V[Z]$$
$$= \sigma^2.$$

Therefore the variance of X is σ^2 and since $\sigma > 0$, its standard deviation is σ.

Let the probability density function and the cumulative distribution function of X be f and F respectively.

$$F(x) = P(X \leqslant x)$$
$$= P(\sigma Z + \mu \leqslant x)$$
$$= P\left(Z \leqslant \frac{x - \mu}{\sigma}\right)$$
$$= \Phi\left(\frac{x - \mu}{\sigma}\right)$$

where Φ is the cumulative distribution function of Z. Differentiating with respect to x, using the chain rule on the right-hand side.

$$F'(x) = \frac{1}{\sigma} \Phi'\left(\frac{x - \mu}{\sigma}\right).$$

But the derivative of a cumulative distribution function is the corresponding probability density function, thus $F' = f$ and $\Phi' = \phi$.

$$f(x) = \frac{1}{\sigma} \phi\left(\frac{x - \mu}{\sigma}\right)$$

$$= \frac{1}{\sigma\sqrt{(2\pi)}} \exp\left(-\frac{(x - \mu)^2}{2\sigma^2}\right).$$

A continuous random variable X with probability density function f given by

$$f(x) = \frac{1}{\sigma\sqrt{(2\pi)}} \exp\left(-\frac{(x - \mu)^2}{2\sigma^2}\right), \qquad -\infty < x < \infty$$

where $\sigma > 0$, is said to be *normally distributed with mean μ and variance σ^2*. This statement is often abbreviated as

$$X \sim N(\mu, \sigma^2).$$

The graph of f:
1 is bell shaped and symmetrical about the line $x = \mu$;
2 is asymptotic to the x-axis, i.e. $f(x) \to 0$ as $x \to \pm\infty$;
3 has two points of inflection where $x = \mu \pm \sigma$;
4 is such that the area of the region bounded by the curve and the x-axis is 1.

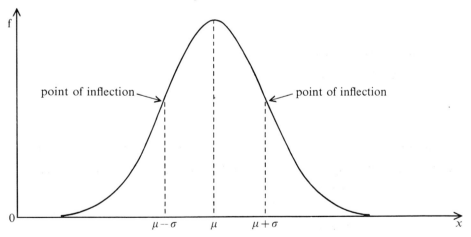

The expression for the normal probability density function

$$f(x) = \frac{1}{\sigma\sqrt{(2\pi)}} \exp\left(-\frac{(x - \mu)^2}{2\sigma^2}\right)$$

contains two parameters, μ and σ, which are, respectively, the mean and standard deviation of the distribution. For each pair of values of μ and σ there is a unique normal distribution.

Suppose that f_1 and f_2 are the probability density functions of two normal random variables having means μ_1, μ_2 and standard deviations σ_1, σ_2 respectively; in addition, suppose that $\mu_1 < \mu_2$ and $\sigma_1 < \sigma_2$. The diagram below shows the graphs of f_1 and f_2 drawn with the same scales and axes. Since $\mu_1 < \mu_2$, the graph of f_1 is located to the left of the graph of f_2 and since $\sigma_1 < \sigma_2$, the graph of f_1 has a higher peak and is less widespread than the graph of f_2.

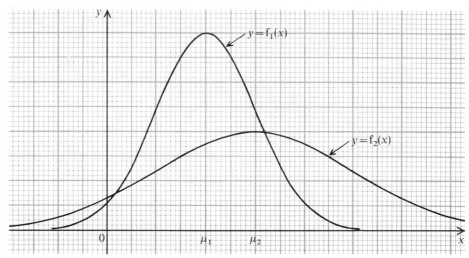

Further use of Table 3

When $X \sim N(\mu, \sigma^2)$, the distribution function of X cannot be expressed in terms of simple functions, but it is possible to calculate probabilities associated with X using the transformation $Z = \dfrac{(X - \mu)}{\sigma}$ in conjunction with **Table 3**, page 217.

In the examples which follow extensive use will be made of the following result.

If $X \sim N(\mu, \sigma^2)$ and $Z = \dfrac{(X - \mu)}{\sigma}$, then $Z \sim N(0, 1)$.

Example 2

Given that $X \sim N(2, 25)$, find

a $P(X < 7)$ **b** $P(X > 6)$ **c** $P(0 < X < 4)$

d $P(X < 0)$ **e** the value of a if $P(X < a) = 0.95$.

$X \sim N(2, 25)$ means that $\mu = 2$ and $\sigma^2 = 25$, thus $\mu = 2$ and $\sigma = 5$

$$Z = \frac{(X - 2)}{5} \sim N(0, 1)$$

a $P(X < 7) = P\left(Z < \frac{(7-2)}{5}\right)$

$\quad\quad\quad = P(Z < 1)$

$\quad\quad\quad = \Phi(1)$

$\quad\quad\quad = 0.8413$

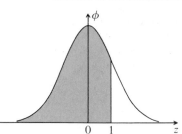

b $P(X > 6) = P\left(Z > \frac{(6-2)}{5}\right)$

$\quad\quad\quad = P(Z > 0.8)$

$\quad\quad\quad = 1 - \Phi(0.8)$

$\quad\quad\quad = 1 - 0.7881$

$\quad\quad\quad = 0.2119$

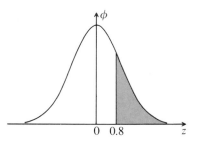

c $P(0 < X < 4) = P\left(\frac{(0-2)}{5} < Z < \frac{(4-2)}{5}\right)$

$\quad\quad\quad\quad\quad = P(-0.4 < Z < 0.4)$

$\quad\quad\quad\quad\quad = \Phi(0.4) - \Phi(-0.4)$

$\quad\quad\quad\quad\quad = \Phi(0.4) - (1 - \Phi(0.4))$

$\quad\quad\quad\quad\quad = 2\Phi(0.4) - 1$

$\quad\quad\quad\quad\quad = 2 \times 0.6554 - 1$

$\quad\quad\quad\quad\quad = 0.3108$

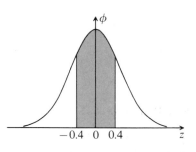

d $P(X < 0) = P(Z < -0.4)$

$\quad\quad\quad = \Phi(-0.4)$

$\quad\quad\quad = 1 - \Phi(0.4)$

$\quad\quad\quad = 1 - 0.6554$

$\quad\quad\quad = 0.3446$

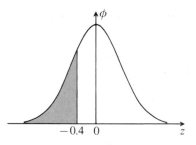

e $\quad\quad\quad P(X < a) = 0.95$

$\quad P\left(Z < \frac{(a-2)}{5}\right) = 0.95$

$\quad\quad \Phi\left(\frac{(a-2)}{5}\right) = 0.95$

since $\Phi(1.64) = 0.9495$ and $\Phi(1.65) = 0.9505$

$\quad\quad \frac{(a-2)}{5} = 1.645$

$\quad\quad\quad\quad a = 1.645 \times 5 + 2$

$\quad\quad\quad\quad a = 10.225$

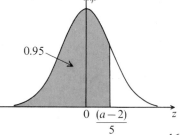

161

Exercise ⌊9.2⌋

1 Given that $X \sim N(15, 16)$, find

  a $P(X < 23)$ **b** $P(16 < X < 19)$ **c** $P(X > 20)$

  d $P(X > 13)$ **e** the value of a if $P(X < a) = 0.975$.

2 Given that $X \sim N(20, 100)$, find

  a $P(X < 26)$ **b** $P(15 < X < 27)$ **c** $P(X > 28)$

  d $P(X < 12)$ **e** the value of a if $P(X < a) = 0.01$.

3 Given that $X \sim N(85, 25)$, find

  a $P(X < 92)$ **b** $P(80 < X < 100)$ **c** $P(X > 96.4)$

  d $P(X < 78.6)$ **e** the value of a if $P(X > a) = 0.005$.

4 Given that $X \sim N(112, 36)$, find

  a $P(X < 124)$ **b** $P(110 < X < 115)$ **c** $P(X > 100)$

  d $P(X > 120)$ **e** the value of a if $P(X > a) = 0.67$.

5 Given that $X \sim N(78, 14)$, find

  a $P(X < 82)$ **b** $P(71 < X < 85)$ **c** $P(X > 84)$

  d $P(X > 73)$ **e** the value of a if $P(X > a) = 0.85$.

9.3 Further worked examples

Example 3

For comparison purposes a headteacher decides that marks in all subjects taken by Form 4 at the end-of-year examinations are to be converted into a form such that the mean and the standard deviation of the converted marks for each subject are 50 and 15 respectively. Given that the raw Mathematics marks are $N(53, 20^2)$ and the raw English marks are $N(49, 10^2)$, calculate the converted marks of a candidate whose raw marks in Mathematics and English are 65 and 57 respectively. Determine whether the candidate would obtain a merit award in either of these subjects if a merit is awarded to the top 25% of the candidates in each subject.

The standardised score of a candidate in a subject is obtained by subtracting the mean score in that subject from the candidate's mark in the subject and then dividing the result by the standard deviation of the marks in the subject. The converted score is obtained by multiplying the standardised score by the selected standard deviation (15 in this example) and adding the selected mean (50 in this example) to the result.

Standardised score in Mathematics $= \dfrac{(65 - 53)}{20}$
$= 0.6$

Converted score in Mathematics $= 0.6 \times 15 + 50$
$= 59$

Standardised score in English $= \dfrac{(57 - 49)}{10}$
$= 0.8$

Converted score in English $= 0.8 \times 15 + 50$
$= 62$

Let m be the standarised score necessary to obtain a merit

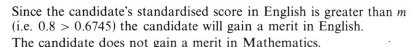

$$P(Z > m) = 0.25$$
$$\Phi(m) = 1 - 0.25$$
$$\Phi(m) = 0.75$$
since $\Phi(0.67) = 0.7486$ and $\Phi(0.68) = 0.7517$

$$m = 0.67 + \frac{0.0014}{0.0031}[0.68 - 0.67]$$
$$m = 0.6745.$$

Since the candidate's standardised score in English is greater than m (i.e. $0.8 > 0.6745$) the candidate will gain a merit in English.
The candidate does not gain a merit in Mathematics.

Example 4

The operational lifetimes of a certain make of light bulb are normally distributed with mean 1100 h and standard deviation 25 h. On a day when the total production of the light bulbs is 10 000, calculate the expected number of light bulbs which have an operational lifetime greater than 1050 h.

The operational lifetime $\quad X \sim N(1100, 25^2)$

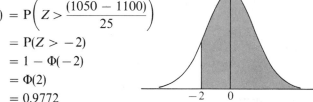

$$Z = \frac{(X - 1100)}{25} \sim N(0, 1)$$

$$P(X > 1050) = P\left(Z > \frac{(1050 - 1100)}{25}\right)$$
$$= P(Z > -2)$$
$$= 1 - \Phi(-2)$$
$$= \Phi(2)$$
$$= 0.9772$$

Let Y be the number of bulbs out of 10 000 having a lifetime greater than 1050
$$Y \sim B(10\,000, 0.9772)$$
$$E[Y] = np = 9772$$

Example 5

A machine produces cylindrical rods whose diameters are normally distributed. A rod is acceptable if its diameter lies between 12.45 cm and 12.55 cm. Records show that 2.5% of the rods produced are rejected because their diameters are too small and 5% of the rods are rejected because their diameters are too large. Calculate, to three decimal places, the mean and standard deviation of the rods produced by the machine.

The diameter $X \sim N(\mu, \sigma^2)$

$$Z = \frac{(X - \mu)}{\sigma} \sim N(0, 1)$$

Since $\quad\quad\quad P(X < 12.45) = 0.025$

$$P\left(Z < \frac{(12.45 - \mu)}{\sigma}\right) = 0.025$$

$$\Phi\left(\frac{(12.45 - \mu)}{\sigma}\right) = 0.025$$

$$\Phi\left(-\frac{(12.45 - \mu)}{\sigma}\right) = 0.975$$

$$-\frac{(12.45 - \mu)}{\sigma} = 1.96$$

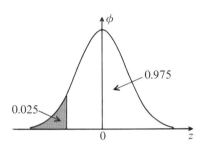

rearranging $\quad\quad 12.45 - \mu = -1.96\sigma \quad\quad\quad (1)$

Since $\quad\quad\quad P(X > 12.55) = 0.05$

$$P\left(Z > \frac{(12.55 - \mu)}{\sigma}\right) = 0.05$$

$$1 - \Phi\left(\frac{(12.55 - \mu)}{\sigma}\right) = 0.05$$

$$\Phi\left(\frac{(12.55 - \mu)}{\sigma}\right) = 0.95$$

$$\frac{(12.55 - \mu)}{\sigma} = 1.645$$

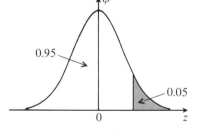

rearranging $\quad\quad 12.55 - \mu = 1.645\sigma \quad\quad\quad (2)$

subtracting (1) from (2) $\quad 0.10 = 3.605\sigma$

giving $\quad\quad \sigma = 0.0277392$ and $\mu = 12.504369$.

Therefore, to three decimal places, the mean and standard deviation are 12.504 cm and 0.028 cm respectively.

Example 6

The diameters of mass-produced spherical marbles are normally distributed with mean 2.5 cm and standard deviation 0.3 cm. Calculate, to two decimal places,

a the upper quartile of the distribution

b the expected value of the surface area of the marbles.

The diameter $X \sim N(2.5, 0.3^2)$

$$Z = \frac{(X - 2.5)}{0.3} \sim N(0, 1)$$

a Let u cm be the upper quartile

$$P(X < u) = 0.75$$

$$P\left(Z < \frac{(u - 2.5)}{0.3}\right) = 0.75$$

$$\Phi\left(\frac{(u - 2.5)}{0.3}\right) = 0.75$$

$$\frac{(u - 2.5)}{0.3} = 0.6745$$

$$u = 2.70$$

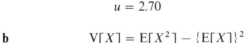

b
$$V[X] = E[X^2] - \{E[X]\}^2$$
$$E[X^2] = V[X] + \{E[X]\}^2$$
$$E[X^2] = 0.3^2 + 2.5^2$$
$$E[X^2] = 6.34$$

Surface area $S = 4\pi R^2 = \pi X^2$

$$E[S] = E[\pi X^2]$$
$$E[S] = \pi \times 6.34$$
$$E[S] = 19.92$$

The expected value of the surface area is $19.92 \, cm^2$.

Exercise 9.3

1 The masses of the contents of cans of beans are normally distributed with mean 235 g and standard deviation 3 g. Calculate the proportion of cans having contents with a mass less than 230 g.

2 The weekly wages paid to employees of a certain firm are approximately normally distributed with mean £117 and standard deviation £15. Calculate the proportion of employees who are paid between £105 and £135.

3 An automatic machine fills packets with sugar. When the machine is set to fill the packets with μ kg of sugar the actual weight of sugar put in the packet is normally distributed with mean μ and standard deviation 0.01 kg.

   a If the machine is set to fill the packets with 1.01 kg of sugar, calculate the proportion of packets which will contain between 0.99 kg and 1.02 kg of sugar.

   b Calculate the value of μ which should be set in order that 99.9% of the packets will contain a weight greater than the nominal weight of 1 kg.

4 The marks obtained in a Mathematics examination are normally distributed with mean 48 and standard deviation 12; these marks are to be converted into a form such that the mean and standard deviation of the converted marks are 50 and 15 respectively. Calculate the converted marks of two candidates whose raw marks are 64 and 40 respectively.

   Given that 70% of the candidates pass and 10% of the candidates obtain a grade A, determine whether the two above-mentioned candidates

   a pass

   b obtain a grade A.

5 A man makes the same journey to work each day. The time at which his work starts is 9.00 a.m. If he leaves home at 8.00 a.m. he is late 5% of the time and if he leaves home at 7.55 a.m. he is late only 1% of the time. Assuming that the times for his journey are normally distributed with the same mean and standard deviation irrespective of the starting times, calculate the mean and standard deviation.

6 Ball-bearings are manufactured with a nominal diameter of 1 cm, but are acceptable if their diameters are inside the limits 0.90 cm and 1.10 cm. It is observed that, in a large batch, 2.5% are rejected as oversize and 2.5% are rejected as undersize. Assuming that the diameters are normally distributed, calculate the percentage rejected if the limits are changed to 0.95 cm and 1.15 cm. Calculate, to three decimal places, the expected value of the surface area of these ball-bearings.

7 The times taken by athletes A and B to run 1500 m are X seconds and Y seconds respectively, where $X \sim N(230, 25)$ and $Y \sim N(240, 100)$. Determine who is more likely to break

   a a track record of 3 minutes 45 seconds

   b a national record of 3 minutes 35 seconds.

8 A company pays a bonus of £10 to any employee who processes in excess of 400 kg of raw material in a day. The daily amount of raw material processed by a certain employee is normally distributed with mean 380 kg and standard deviation 16 kg. Calculate, to the nearest penny, the expected value of the daily bonus of this employee.

9   A soft-drinks machine is regulated so as to deliver an average of 200 ml per cup. The actual amount delivered per cup is a normally distributed random variable having mean 200 ml and standard deviation 8 ml.

   (i) Given that the maximum capacity of a cup is 210 ml find, to three decimal places, the probability that a cup will overflow.

   (ii) Find, to the nearest ml, what the maximum capacity of each cup should be for only 1% of the cups to overflow.

$(WJEC)$

10   Mass-produced right-circular cylindrical pipes have internal diameters that are normally distributed with mean 10 cm and standard deviation of 0.4 cm.

   (i) Find the probability that a randomly chosen pipe will have an internal diameter greater than 10.3 cm.

   (ii) Find the expected number of pipes in a random sample of 100 pipes that have a diameter in the range from 9.7 cm to 10.3 cm.

   (iii) Find the expected value of the internal cross-sectional area of a randomly chosen pipe; give your answer correct to three significant figures.

$(WJEC)$

11   A certain ingredient may be extracted from raw material by either one of two methods, A and B. For a fixed volume of raw material, the amount X cm^3 of the ingredient extracted using method A is normally distributed with mean 13 and standard deviation 2, and the amount Y cm^3 extracted by using method B is distributed with probability density function

$$g(y) = \frac{2}{25}(y - 10), \qquad\qquad 10 \leqslant y \leqslant 15,$$

$$g(y) = 0, \qquad\qquad\qquad\qquad \text{otherwise.}$$

Determine which of the two methods

   (i) has the greater probability of extracting more than 14 cm^3 of the ingredient,

   (ii) extracts, on average, the greater amount of the ingredient.

Suppose that the cost of applying method A is 3p per cm^3 extracted and that the cost of extracting Y cm^3 using method B is $(15 + 2Y)$p. If the extracted ingredient is sold at 5p per cm^3, determine which of the two methods gives a higher expected profit per fixed volume of raw material.

$(WJEC)$

9.4 Normal approximation to the binomial distribution

Given that $X \sim B(n, p)$, where n is large and neither p nor $(1 - p)$ is very small, it may be shown that the distribution of X may be approximated by the normal distribution of a random variable Y having mean np and variance $np(1 - p)$. The formal justification of this statement is too difficult at this stage, but the diagrams below indicate graphically that the approximation is only moderately good for $n = 20$ and $p = 0.3$ and very good for $n = 100$ and $p = 0.5$. Each of the diagrams shows a theoretical relative frequency histogram for $B(n, p)$ with the graph of the probability density function of $N(np, np(1 - p))$ superimposed on it.

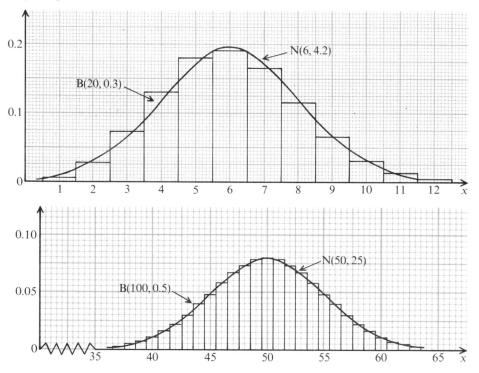

One problem immediately arises from the fact that the binomial distribution is discrete and the normal distribution is continuous. It is clearly impractical to use $P(Y = r)$ as an approximate value for $P(X = r)$, where r is an integer between 0 and n, since the former is zero and the latter non-zero. This is overcome by using $P(r - 0.5 < Y < r + 0.5)$ as an approximate value for $P(X = r)$.

Continuity corrections

When X is a discrete random variable whose range space is a subset of the set of integers and Y is a continuous random variable whose distribution is to be used as an approximation to the distribution of X,

$$P(X = r) \simeq P(r - 0.5 < Y < r + 0.5),$$

$$P(X \leqslant r) \simeq P(Y < r + 0.5),$$

$$P(X \geqslant r) \simeq P(Y > r - 0.5),$$

these are called *continuity corrections*.

The next problem is how to determine when n is large enough and when p is not too small. This is not easy because the values of n and p interact.

When $p = 0.5$ the binomial distribution is symmetrical as is the normal distribution and the use of the normal approximation is reasonable for values of n as small as 20.

When $p \neq 0.5$ the binomial distribution is not symmetrical and it becomes more unsymmetrical the further p deviates from 0.5; however, for each value of p, the distribution becomes more symmetrical as n increases. The further p deviates from 0.5 the larger the value of n must be for the use of a normal approximation to be appropriate.

It is difficult to state in simple terms the conditions under which the use of a normal approximation to a binomial distribution is appropriate, so various rules-of-thumb are adopted. Three alternative rules are listed below:

A normal approximation to $B(n, p)$ may be used when:

Rule 1 $0.1 < p < 0.9$, $np > 5$ and $n(1 - p) > 5$

Rule 2 $np > 10$ and $n(1 - p) > 10$

Rule 3 n is greater than the bigger of $\dfrac{16p}{(1 - p)}$ and $\dfrac{16(1 - p)}{p}$.

The list is not exhaustive. Rule 1 is usually the least strict, whilst Rule 3 is usually the most strict and the easiest to justify.

$$\text{Let } X \sim B(n, p) \text{ and } Y \sim N(np, np(1 - p)).$$

The values of X range from 0 to n, whilst the values of Y range from $-\infty$ to $+\infty$. For a good approximation it is necessary that $P(0 < Y < n)$ should be as near to 1 as possible. Since the probability that a normal variable lies between $\mu - 4\sigma$ and $\mu + 4\sigma$ is 0.99994 it is desirable that

$$np - 4\sqrt{np(1 - p)} > 0 \quad \text{and} \quad np + 4\sqrt{np(1 - p)} < n$$

$$np > 4\sqrt{np(1 - p)} \qquad \text{and} \qquad 4\sqrt{np(1 - p)} < n(1 - p)$$

squaring:

$$n^2 p^2 > 16np(1 - p) \quad \text{and} \quad 16np(1 - p) < n^2(1 - p)^2$$

$$n > \frac{16(1 - p)}{p} \qquad \text{and} \qquad \frac{16p}{(1 - p)} < n.$$

Thus justifying the use of Rule 3.

Example 7

Given that $X \sim B(100, 0.4)$, use a normal approximation to evaluate

a $P(X = 40)$ **b** $P(X \geq 50)$ **c** $P(X < 35)$.

Since $X \sim B(100, 0.4)$, $\mu = np = 40$ and $\sigma^2 = np(1 - p) = 24$.

Let $Y \sim N(40, 24)$ then $Z = \dfrac{Y - 40}{\sqrt{24}} \sim N(0, 1)$

a $P(X = 40) \simeq P(39.5 < Y < 40.5)$

$\simeq P\left(\dfrac{39.5 - 40}{\sqrt{24}} < Z < \dfrac{40.5 - 40}{\sqrt{24}}\right)$

$\simeq P(-0.102 < Z < 0.102)$

$\simeq \Phi(0.102) - \Phi(-0.102)$

$\simeq 0.5406 - (1 - 0.5406)$

$\simeq 0.0812$

This is exactly the same as the value (to 4 d.p.) calculated directly from the binomial distribution.

b $P(X \geq 50) \simeq P(Y > 49.5)$

$\simeq P\left(Z > \dfrac{49.5 - 40}{\sqrt{24}}\right)$

$\simeq P(Z > 1.939)$

$\simeq 1 - \Phi(1.939)$

$\simeq 1 - 0.9737$

$\simeq 0.0263$

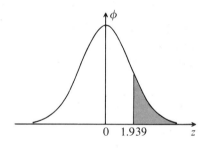

This compares with the answer 0.0271 (to 4 d.p.) obtained by a direct calculation.

c $P(X < 35) \simeq P(Y < 34.5)$

$\simeq P\left(Z < \dfrac{34.5 - 40}{\sqrt{24}}\right)$

$\simeq P(Z < -1.123)$

$\simeq 1 - \Phi(1.123)$

$\simeq 1 - 0.8692$

$\simeq 0.1308$

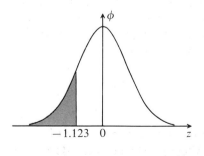

This compares with the answer 0.1303 (to 4 d.p.) obtained by a direct calculation.

Example 8

Independently for each fly sprayed with a certain spray the probability that the fly will be killed is 0.7.

a A sample of 100 flies is sprayed. Calculate an approximate value for the probability that at least 65 of the flies sprayed are killed.

b Estimate the least sample size for which the probability that at least 65% of the flies sprayed will be killed is greater than 0.9.

a Let X be the number of flies killed in the sample.

$$X \sim B(100, 0.7), \quad \mu = np = 70 \text{ and } \sigma^2 = np(1 - p) = 21.$$

Let $Y \sim N(70, 21)$ then $Z = \dfrac{Y - 70}{\sqrt{21}} \sim N(0, 1)$.

$P(X \geqslant 65) \simeq P(Y > 64.5)$

$$\simeq P\left(Z > \frac{64.5 - 70}{\sqrt{21}}\right)$$

$$\simeq P(Z > -1.200)$$

$$\simeq \Phi(1.200)$$

$$\simeq 0.8849$$

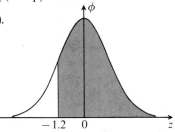

b Let X be the number of flies killed in the sample, of size n.

$$X \sim B(n, 0.7), \quad \mu = np = 0.7n \text{ and } \sigma^2 = np(1 - p) = 0.21n.$$

Let $Y \sim N(0.7n, 0.21n)$, then $Z = \dfrac{(Y - 0.7n)}{\sqrt{(0.21n)}} \sim N(0, 1)$.

$P(X \geqslant 0.65n) > 0.9$

$P(Y > 0.65n - 0.5) > 0.9$

$$P\left(Z > \frac{0.65n - 0.5 - 0.7n}{\sqrt{(0.21n)}}\right) > 0.9$$

$$P\left(Z > \frac{-0.05n - 0.5}{\sqrt{(0.21n)}}\right) > 0.9$$

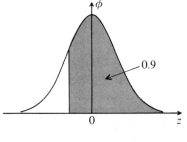

$$\frac{-0.05n - 0.5}{\sqrt{(0.21n)}} \simeq -1.282 \qquad (1)$$

This equation may be solved by squaring both sides and forming a quadratic in n; this gives an answer of 118 for n.

Another less accurate method is to neglect the continuity correction; in this case the equation becomes

$$\frac{0.05n}{\sqrt{(0.21n)}} \simeq 1.282$$

$$\sqrt{n} \simeq \frac{1.282 \times \sqrt{0.21}}{0.05}.$$

Giving an estimate for n of 139.

Alternatively, equation (1) may be solved by trial after finding an approximate value by the less accurate method described above. A trial value of 120 indicates that 120 is too great and the value of 118 is soon found.

Exercise 9.4

1 Given that $X \sim B(100, 0.3)$, calculate approximate values for
  a $P(X = 32)$ **b** $P(X \leqslant 35)$ **c** $P(X > 25)$.

2 Given that $X \sim B(50, 0.5)$, calculate approximate values for
  a $P(X = 25)$ **b** $P(X \geqslant 30)$ **c** $P(X < 27)$.

3 Given that $X \sim B(400, 0.2)$, calculate approximate values for
  a $P(X = 84)$ **b** $P(75 < X < 85)$ **c** $P(X \geqslant 85)$.

4 Two fair coins are tossed together 100 times. Find an approximate value for the probability that two heads will be obtained at least 20 times.

5 An electronic device consists of n components and it will function for a given period of time only if at least 75% of these components function for that period. If, independently for each component, the probability that the component will function for the given period is 0.8, estimate the least value of n such that there is a probability of at least 0.95 that the device will function throughout the period.

Miscellaneous Exercise 9

1 A fair coin is to be tossed 100 times. Find an approximate value for the probability that at least 60 heads will be obtained. *(JMB)*

2 In each trial of a certain experiment the probability of an event A occurring is 0.6. Find the probabilities, correct to three decimal places, that in five independent trials of the experiment, A will occur
(i) less than three times,   (ii) at least four times.

Use a suitable approximation to evaluate, correct to three decimal places, the probability that A occurs at least 69 times in 100 independent trials of the experiment. *(JMB)*

3 A complex electronics system consists of 100 components which function independently. The probability that a component will fail during a period of operation of the system is 0.2. The entire system will function only when at least 75 of the components are functioning. Using the normal approximation to a binomial distribution, estimate the probability that the system will function during the period of operation. *(JMB)*

4 Observation of a very large number of cars at a certain point on a motorway establishes that speeds are normally distributed. 90% of cars have speeds less than 75.5 mph and only 5% have speeds less than 60 mph. Determine the mean speed μ and the standard deviation σ. Give your answers correct to two decimal places.

Because of a fuel economy drive the mean speed of the motorists is reduced. Assuming that the standard deviation σ has remained unchanged and that 2% now exceed 75 mph, find the new mean speed. What percentage of the motorists now exceed the 70 mph speed limit? *(WJEC)*

5 a Steel rods are required to be at least 20 cm long and a machine cuts them with a mean length of 20.04 cm and a standard deviation of 0.03 cm. Assuming a normal probability density, determine what proportion of the rods will be rejected for being too short.

    (i) If the spread of the measurements is unaltered by adjusting the position of the cut, how large should the mean be made if only a 1% rejection rate can be tolerated? Give your answer to 2 decimal places.

    (ii) Alternatively, to what would the standard deviation have to be reduced if the mean length was to be unaltered but a 1% rejection rate achieved? Give your answer to 2 decimal places.

b A fair coin is tossed 100 times. Use the normal approximation to the binomial probability function to find the probability of getting more than 53 heads.

Repeat the calculation to find the probability of getting more than 530 heads in 1000 tosses.

            (*WJEC*)

6 Describe conditions under which it is appropriate to approximate a binomial distribution by a normal distribution.

In each trial of a random experiment it is known that the event A has a probability 0.58 of occurring. In 100 independent trials of the experiment, let X denote the number of times that A occurs. Write down an exact expression for the probability $P(X \geqslant 50)$, but do not evaluate it. Use the normal approximation to find an approximate value for this probability.

            (*JMB*)

7 The operational life-times of certain electronic components are found to be normally distributed with mean 5200 hours and standard deviation 400 hours.

Calculate, to three significant figures,

    (i) the proportion of such components having life-times between 4500 and 5800 hours,

    (ii) the 67th percentile of this distribution,

    (iii) the probability that when three components are chosen at random none of them will have a life-time within 400 hours of the mean.   (*JMB*)

8 The lengths of metal bars are normally distributed with a mean of 100 cm and a standard deviation of 2 cm. A bar of length less than 98 cm is rejected; a bar of length between 98 cm and $(100 + c)$ cm, where c is a positive constant, is filed down to a length of 98 cm; a bar of length greater than $(100 + c)$ cm has a piece of length $(2 + c)$ cm cut from it and it is then filed down to a length of 98 cm.

    (i) Find the proportion of bars rejected.

    (ii) If $c = 3$, find the proportion of non-rejected bars that have to be cut.

    (iii) Find the value of c if the proportion of non-rejected bars that have to be cut is 0.01.   (*JMB*)

9 An automatic device is used for cutting steel strips into standard lengths. When the device is set to cut lengths of μ metres it is known that the actual lengths cut are normally distributed with mean μ and a standard deviation of σ metres, σ being constant for all μ. It is also known that 90 per cent of the cut lengths are within 0.002 metres of the length set on the device.

   (i) Calculate the value σ.

  (ii) Calculate the probability that exactly two of five randomly chosen strips will have lengths within 0.002 metres of the length set on the device.

 (iii) If the setting on the device shifts from μ to $\mu + 0.001$, calculate the proportion of the cut lengths that are within 0.002 metres of the old setting.

(JMB)

10 In a given manufacturing process, components are rejected if they have a particular dimension greater than 60.4 mm or less than 59.7 mm. It is found that 3% are rejected as being too large and 5% are rejected as being too small. Assuming that the dimension is normally distributed, find the mean and standard deviation of the distribution of the dimension, correct to one decimal place.

Use the mean and standard deviation you have calculated to estimate the percentage of rejects if the limits for acceptance are changed to 59.6 mm and 60.3 mm.

(WJEC)

11 A salesman makes a journey from his home to a given supermarket on the same day every week. He wants to be at the supermarket when it opens at 10.00 a.m. When he leaves home at 8.20 a.m. he is late once in twenty times, whereas when he leaves home at 8.15 a.m. he is late once in a hundred times.

   (i) Assuming that the time taken to make the journey has the same normal distribution irrespective of his starting time, at what time should he leave home in order to be late not more than once in two hundred times?

  (ii) For a continuous period of thirty weeks, roadworks on the salesman's route cause the mean time for his journey to be increased by 12 minutes, the standard deviation remaining the same and the distribution remaining normal. Given that he leaves home during this 30-week period at 8.15 a.m., on how many occasions would you expect him to be late? Give your answer correct to the nearest whole number.

(WJEC)

12 A certain production department of a large engineering works produces items with dimensions 15.00 ± 0.05 mm, and every article produced is inspected to see whether it satisfies these limits. The dimension is normally distributed but there is a bias in the machine setting so that the mean dimension of the articles produced is 15.01 mm. The standard deviation of the dimension is 0.05 mm and it costs 20p to produce each article.

a 1000 articles per month are produced. Find the expected profit per month if the non-defective articles sell for 30p each, with those outside the tolerance limits being sold as rejects at 2p each.

b In order to increase the expected profit, the standard deviation was reduced to 0.04 mm at an extra production cost of 2p per article. Calculate the new expected profit. *(WJEC)*

13 A boy would like to complete his homework by 9 o'clock each evening from Monday to Friday. He decided to start at 7 o'clock and over a period of time found that he finished after 9 o'clock on average once each week. He decided to start at ten minutes to seven instead, and then found that he finished after 9 o'clock on average once every four weeks.

Assuming that the durations of his homework are independent and normally distributed, find the time he should start if he does not want to work after 9 o'clock more than once a term. (1 term = 15 weeks.) *(JMB)*

14 Define the binomial distribution, stating clearly the conditions under which it applies. State also the conditions under which it can be approximated by a normal distribution.

(i) In each of ten sets of twins, one member is chosen at random and given diet A and the other is given diet B. After a certain time their gains in weight are compared. If diet A is neither more nor less effective than diet B, calculate the probability that eight or more of the twins given diet A gain more weight than their partners.

(ii) If 55 per cent of the adults in a large city are male, determine the probability that a random sample of 400 adults will contain more female than male members. *(JMB)*

15 Derive the formulae for the mean and variance of the binomial distribution. Given that the probability that a child is left-handed is $\frac{1}{5}$, calculate the probabilities that in a random sample of five children,

(i) exactly 2 are left-handed.

(ii) more than 2 are left-handed.

There are 1600 children in a school. Use the normal approximation to calculate the probability that the school contains between 330 and 350 (inclusive) left-handed children. *(JMB)*

16 Mass-produced metal spheres have diameters which are normally distributed with mean 15 cm and standard deviation 0.26 cm.

(i) Find, to three decimal places, the probability that the diameter of a randomly chosen sphere lies between 14.72 cm and 15.48 cm.

(ii) Given that 95 per cent of all the spheres have diameters exceeding d cm, find the value of d to two decimal places.

(iii) Calculate, to one decimal place, the expected value of the surface area of a randomly chosen sphere.
(The surface area of a sphere of diameter d is πd^2.) *(JMB)*

17 The continuous random variable X is uniformly distributed between 9 and 21. The independent random variable Y has a normal distribution with the same mean and variance as X. Find, to three decimal places,

(i) the probability that X and Y take values both of which are less than 18,

(ii) the probability that X and Y take values one of which is less than 18 and one which is greater than 18.   (*JMB*)

18 Write down the probability of r successes in n independent trials with constant probability of success p. Determine the mean value of r. Write down the variance of r. State the circumstances in which it is possible to approximate to the probability distribution for r by means of the normal distribution.

The probability that a greenfly will be killed by a certain insect spray is 0.8. A sample of 100 greenfly is sprayed. Use the normal distribution as an approximation to the probability distribution for the number of greenfly killed to find the probability that at least 75 are killed by the spray.

For a sample of greenfly of size n, state the mean and variance of the proportion that will be killed by the spray. Find the least sample size for which there is a greater than 95% probability that at least 75% will be killed by the spray.   (*JMB*)

19 State conditions under which it is permissible to use the normal distribution as an approximation to the binomial distribution.

In a multiple-choice examination paper there are 100 questions, each with a choice of three answers of which only one is correct. To pass the examination it is necessary to answer 40 or more questions correctly. Use the normal distribution as an approximation to the binomial distribution to estimate the probability that a candidate who chooses the answer to each question randomly will pass the examination.

Estimate the probability of passing in this way if the choice of answers to each question is increased to four.

Assuming that the choice of answers to each question remains at three, and the proportion of correct choices required to pass remains at 40%, estimate the least number of questions that the paper should contain if the probability of a pass by random choice is not to exceed 1%.   (*JMB*)

20 Seeds of a certain kind are planted in plastic strips, each strip containing 20 seeds. The strips are arranged on trays, each tray containing six strips, and then left to germinate. The seeds germinate independently, the probability of any particular seed germinating being 0.9. Show that the probability that in any one tray at least five of the six strips will each contain more than 17 germinated seeds is 0.372 correct to three decimal places.

Fifty randomly selected trays are examined. Find an approximate value for the probability that at least 25 of the 50 trays have at least five strips each containing more than 17 germinated seeds, giving your answer to two decimal places.   (*JMB*)

Chapter 10

Revision

This chapter contains brief notes and reminders about important points, followed by exercises.

10.1 Descriptive statistics

Frequency distributions

A frequency distribution derived from data concerning a discrete variable is usually illustrated by means of a (vertical) line diagram which emphasises the discrete nature of the variable. The associated cumulative frequency distribution is illustrated by means of a graph of a step function.

Grouped frequency distributions

A grouped frequency distribution is usually illustrated by means of a histogram. The following points concerning histograms should be noted:

1 The base of each rectangle in a histogram extends from the lower class boundary to the upper class boundary of the corresponding class interval.
2 The height of each rectangle is calculated by dividing the class frequency by the class width.
3 The area of each rectangle represents the class frequency.
4 The 'vertical' axis must start at zero.
5 The 'vertical axis' should be labelled 'frequency/class width' or 'frequency density'; it should not be labelled 'frequency'.

In a relative frequency histogram the area of each rectangle represents the relative frequency of the class and the total area of all the rectangles is unity. A relative frequency histogram is identical to the corresponding histogram apart from the scale on the 'vertical' axis.

The cumulative frequency distribution associated with a grouped frequency distribution is illustrated by means of a cumulative frequency polygon or a cumulative frequency curve (ogive). The following points should be noted:

1 The cumulative frequency of each class is plotted against the upper class boundary of that class.

2 Successive points are joined by straight lines when the diagram is to be used to estimate a summary measure of the data set.

3 Successive points are joined with a smooth curve when the diagram is to be used to estimate a summary measure of the population from which the data set was obtained.

Mean, variance and standard deviation

The most commonly used summary measure of location is the (arithmetic) mean.

The mean
$$\bar{x} = \frac{\sum\limits_{i=1}^{k} f_i x_i}{n},$$
where $n = \sum\limits_{i=1}^{k} f_i.$

The measure of dispersion used in conjunction with the mean is the variance or, alternatively, the standard deviation which is the positive square root of the variance.

The variance
$$\text{var}(x) = \frac{\sum\limits_{i=1}^{k} f_i(x_i - \bar{x})^2}{n}$$
or
$$\text{var}(x) = \frac{\sum\limits_{i=1}^{k} f_i x_i^2}{n} - \bar{x}^2.$$

If $y_i = ax_i + b$, where a and b are constants,

then
$$\bar{y} = a\bar{x} + b$$

and
$$\text{var}(y) = a^2 \text{var}(x).$$

The mean, variance and standard deviation take into account every value in the data set; this is usually an advantage but occasionally they may be unduly distorted by atypical values.

Median, percentiles and interquartile range

The median of a data set is the value of the middle observation (or the mean of the values of the two middle observations) when the observations are arranged in order of magnitude.

The kth percentile is a value associated with the cumulative frequency $\frac{kn}{100}$ in accordance with the following rules:

1 If $\frac{kn}{100}$ is an integer r, then the kth percentile is the mean of the rth and $(r + 1)$th observations in the ordered data set.

2 If $\frac{kn}{100}$ is not an integer but lies between the integers r and $(r + 1)$, then the kth percentile is the $(r + 1)$th observation in the ordered data set.

The 50th percentile is the median, whilst the 25th and 75th percentiles are called the lower quartile and the upper quartile respectively. The difference between the upper and lower quartiles, called the interquartile range, is often used as a measure of dispersion when the median is used as a measure of location.

The kth percentile of a distribution may be found from a cumulative frequency diagram by reading the value of the variable corresponding to the cumulative frequency of $\dfrac{kn}{100}$.

Exercise 10.1

1 Given the set of numbers 2, 4, 4, 5, 6, 6, 8, 10, 11, 12, 13, 15, 17, 41,
   a find the median and interquartile range
   b find the mean and standard deviation
   c state, with a reason, which measure of location is preferable.

2 The following table gives the distribution of the number of defective items in 100 boxes, each containing 144 items.

Number of defective items	0	1	2	3	4	5
Frequency	22	37	20	13	6	2

   a Illustrate this frequency distribution.
   b Find the mean and variance.
   c Find the median.

3 A firm employs 5 salespersons in each of two showrooms. One week, the mean and variance of the number of cars sold per employee in one showroom were 6 and 6 respectively and the corresponding results in the second showroom were 8 and 6 respectively. Find the mean and variance of the number of cars sold per employee for the combined group of 10 salespersons.

   As the salespersons had exceeded the firm's sales target for the week, each salesperson was given a bonus of £20 plus £10 for each car sold by that person. Find the mean and standard deviation of the bonus payments paid to the 10 salespersons during the week.

4 In a certain area, the distribution of the ages (in completed years) of 600 children under 11 years old is given in the following table.

Age (in completed years)	0–2	3–4	5–6	7	8	9	10
Number of children	189	124	114	50	44	42	37

   a Draw a histogram to illustrate this distribution.
   b Estimate the mean and standard deviation of the ages of these children.
   c Estimate the median age of these children.
   d Given that there are 400 children aged 11 or over in the area, estimate the median age of all 1000 children in the area.

5 A tyre manufacturer conducts certain trials on a particular type of tyre. A random sample of 100 tyres is put on test and the distances travelled by the tyres before reaching the legal limit of tyre wear are shown in the following table.

Distance in km	Number of tyres
5000 to 25000	8
25000 to 35000	14
35000 to 45000	24
45000 to 55000	26
55000 to 65000	16
65000 to 85000	12

(i) Plot these data as a histogram. (Note the range of the first and last class intervals.)

(ii) Calculate the mean distance, showing all your working.

(iii) Obtain the cumulative frequency distribution and estimate the median and interquartile range.

These data were collected by fitting the tyres to the front wheels of a fleet of hire cars of the same model. Suppose that the manufacturer had, instead, tested the tyres by running them on constant speed rollers in a simulated wear trial. Describe the effect you think this would have had on the distribution of the distances travelled by the tyres.

Discuss briefly the relative merits of the two methods for examining tyre wear. *(JMB)*

10.2 Probability

It is essential to learn the rules of probability listed below.

Rule 1 $P(A') = 1 - P(A)$.

Rule 2 $P(A \cup B) = P(A) + P(B) - P(A \cap B)$.

Rule 3 If A and B are mutually exclusive, then $P(A \cup B) = P(A) + P(B)$.

Rule 4 $P(A \cap B') = P(A) - P(A \cap B)$.

Rule 5 $P(B \mid A) = \dfrac{P(A \cap B)}{P(A)}$ or $P(A \cap B) = P(A) . P(B \mid A)$.

Rule 6 $P(A \cap B) = P(A) . P(B)$ if and only if A and B are independent.

Rule 7 If $\{A_1, A_2, \ldots, A_n\}$ is a set of mutually exclusive and exhaustive events then
$$P(B) = P(A_1) . P(B \mid A_1) + P(A_2) . P(B \mid A_2) + \ldots + P(A_n) . P(B \mid A_n).$$

Permutations and combinations

The following results are often very useful in questions involving probability.

The number of ways of arranging n objects in a row, when there are p alike of one kind, q alike of a second kind and r alike of a third kind, is

$$\frac{n!}{p!\,q!\,r!}.$$

The number of ways of selecting r objects from n different objects when the order of selection is not important is

$$\binom{n}{r} = \frac{n!}{r!\,(n-r)!}.$$

Exercise 10.2

1 Three events A, B, C are such that

$$P(A) = 0.5, \qquad P(B) = 0.6, \qquad P(C) = 0.2.$$

A and C are independent, B and C are mutually exclusive and the probability that only one of these three events will occur is 0.5. Show that A and B are independent.

2 Three boxes A, B, C are identical in external appearance. Box A contains 5 red, 3 white and 2 green balls; box B contains 4 red, 3 white and 3 green balls; box C contains 3 red, 1 white and 6 green balls. One box is chosen at random and three balls are drawn at random without replacement from it.

a Calculate the probability that one ball of each colour is drawn.

b Given that one ball of each colour was drawn, calculate the probability that they were drawn from box A.

3 In a target pistol shooting contest two men A and B compete to see who can hit the bull first. Independently, the probabilities that A and B will hit the bull on each shot are 0.2 and 0.1 respectively. They take alternate shots with B starting first. Find the probability that A will win the contest.

In a practice session, assuming the probability that B will hit the bull on each shot remains at 0.1, find the least number of shots that B must have so that the probability of his hitting the bull at least once is greater than 0.9.

4 A party of 30 people, including A, B and C, is to make a journey in three vehicles with passenger capacities of 8, 10 and 12. The members of the party take their places at random.

Find the probabilities that
(i) A travels in the 8-passenger vehicle,
(ii) B travels in the 8-passenger vehicle,
(iii) both A and B travel in the 8-passenger vehicle,

(iv) A, B and C all travel in the 8-passenger vehicle,

(v) A, B and C all travel in the same vehicle,

(vi) A, B and C travel in different vehicles,

(vii) two of A, B and C travel in one vehicle and the third in a different vehicle. *(JMB)*

5 A census of married couples showed that 50% of the couples had no car, 40% had one car and the remaining 10% had two cars. Three of the married couples are chosen at random.

(i) Find the probability that one couple has no car, one couple has one car and one has two cars.

(ii) Find the probability that the three couples have a combined total of three cars.

The census also showed that both the husband and the wife were in full-time employment in 16% of those couples having no car, in 45% of those having one car and in 60% of those having two cars.

(iii) For a randomly chosen married couple find the probability that both husband and wife are in full-time employment.

(iv) Given that a randomly chosen married couple is one where both husband and wife are in full-time employment, find the conditional probability that the couple has no car. *(JMB)*

10.3 Discrete random variables

A random variable is a function whose domain is the sample space of a random experiment and whose range is a set of real numbers (called the range space of the random variable). If the number of elements in the range space is finite or countably infinite, the random variable is said to be discrete.

A function which assigns a probability to each element of the range space is called the probability function of the random variable.

Summary measures

$$\text{Mean } \mu = \sum_{i=1}^{k} p_i x_i, \qquad \text{where } p_i = P(X = x_i)$$

$$\text{Variance } \sigma^2 = \sum_{i=1}^{k} p_i (x_i - \mu)^2.$$

Alternatively, $\quad\text{Variance } \sigma^2 = \sum_{i=1}^{k} p_i x_i^2 - \mu^2.$

Expected value (or expectation)

$$E[g(x)] = \sum_{i=1}^{k} p_i g(x_i)$$

$$\mu = E[X]$$

$$\sigma^2 = E[X^2] - \{E[X]\}^2$$

$$E[aX + b] = aE[X] + b$$

$$V[aX + b] = a^2 V[X]$$

$$E[aX + bY] = aE[X] + bE[Y]$$

Probability generating functions

$$G(t) = \sum_{r=0}^{\infty} p_r t^r \qquad \text{where } p_r = P(X = r)$$

$$\mu = G'(1)$$

$$\sigma^2 = G''(1) + \mu - \mu^2$$

Binomial distribution

If X is the number of successes in n independent Bernoulli trials, in each of which the probability of a success is a constant p and the probability of a failure is $q = 1 - p$, then

$$P(X = r) = \binom{n}{r} q^{n-r} p^r \qquad r = 0, 1, 2, \ldots, n$$

$$G(t) = (q + pt)^n$$

$$\mu = np$$

$$\sigma^2 = npq$$

Recursive formula $\qquad \dfrac{p_r}{p_{r-1}} = \dfrac{(n - r + 1)p}{rq} \qquad \text{where } p_r = P(X = r).$

Mean and variance of a proportion

If $\hat{P}$ is the observed proportion of successes in n independent Bernoulli trials, in each of which the probability of a success is a constant p, then

$$E[\hat{P}] = p$$

$$V[\hat{P}] = \frac{p(1 - p)}{n}.$$

Discrete uniform distribution

If a discrete random variable X with a finite range space R_x is such that for all $x \in R_x$, $P(X = x) = k$, where k is a constant, then X is said to have a discrete uniform distribution.

Exercise 10.3

1 A bag contains 4 red balls and 8 white balls.

   a If 6 balls are drawn from the bag at random without replacement, find the mean and the variance of the number of red balls drawn.

   Deduce the mean and the variance of the difference between the numbers of white and red balls drawn.

   b If 6 balls are drawn from the bag at random with replacement, find the mean and the variance of the number of red balls drawn.

2 A golfer decides to practise 20 m putts from the same position on a certain green. Suppose that, independently for each putt, the probability of his sinking the putt is 0.1 and that X is the number of attempts he makes up to and including his first success.

   a Find $P(X > 3)$ and $E[X]$.

   b If Y minutes is the time taken by the golfer over the practice session and $Y = 2X + 5$, find $E[Y]$.

3 A discrete random variable has a uniform distribution given by

$$P(X = x) = \frac{1}{n}, \qquad x = 1, 3, 5, \ldots, (2n - 1).$$

   Find the mean and the variance of X.

4 Electronic components of a certain type are mass-produced. If, in a random sample of 50 of these components, 12 are found to be defective, find an estimate for the proportion of the mass-produced components which are defective. Find also an approximate value for the standard error of this estimate.

5 In each of the following $X \sim B(n, p)$.

   a When $n = 20$ and $p = 0.3$, find $P(X < 6)$.

   b When $n = 10$ and $p = 0.8$, find $P(X < 7)$.

   c When $n = 60$ and $p = 0.6$, find $P(X = 36)$.

   d When $n = 50$ and $p = 0.75$, find the mode of the distribution.

   e When $p = 0.18$ and $P(X > 0) > 0.95$, find the least possible value of n.

10.4 Continuous random variables

A continuous random variable is a random variable whose range space is neither finite nor countably infinite.

Probability density function

If X is a continuous random variable with range space $R = \{x : a \leqslant x \leqslant b\}$ and f is a function with domain R such that

$$f(x) \geqslant 0$$

$$\int_a^b f(x)\,dx = 1$$

$$\int_c^d f(x)\,dx = P(c \leqslant X \leqslant d)$$

for all c, d such that $a \leqslant c < d \leqslant b$, then f is called the probability density function of X.

Summary measures

$$\text{Mean } \mu = \int_a^b x f(x)\,dx$$

$$\text{Variance } \sigma^2 = \int_a^b (x - \mu)^2 f(x)\,dx.$$

Alternatively, $$\text{Variance } \sigma^2 = \int_a^b x^2 f(x)\,dx - \mu^2$$

Expected value (or expectation)

$$E[g(X)] = \int_a^b g(x) f(x)\,dx$$

$$\mu = E[X]$$

$$\sigma^2 = E[X^2] - \{E[X]\}^2$$

$$E[aX + b] = aE[X] + b$$

$$V[aX + b] = a^2 V[X]$$

$$E[aX + bY] = aE[X] + bE[Y]$$

$$E[ag(X) + bh(Y)] = aE[g(X)] + bE[h(Y)]$$

Distribution function

The (cumulative) distribution function F is such that
$$F(x) = P(X \leqslant x).$$

Alternatively, $F(x) = 0$ $x < a$

$$F(x) = \int_a^x f(t)\,dt \qquad a \leqslant x \leqslant b$$

$F(x) = 1$ $x > a$

Also $f(x) = F'(x).$

Median and percentiles

The median m of a continuous distribution is found by solving the equation
$$F(m) = 0.5.$$

The kth percentile x_k is found by solving the equation
$$F(x_k) = \frac{k}{100}.$$

Continuous uniform distribution (or rectangular distribution)

The continuous uniform distribution $U(a, b)$ has the probability density function f given by

$$f(x) = \frac{1}{(b-a)}, \qquad a \leqslant x \leqslant b,$$

$$f(x) = 0, \qquad \text{otherwise.}$$

The mean and variance of $U(a, b)$ are $\dfrac{(b+a)}{2}$ and $\dfrac{(b-a)^2}{12}$ respectively.

Exercise 10.4

1 A continuous random variable X has a probability density function f given by
$$f(x) = \frac{1}{\pi(1 + x^2)}, \qquad -\infty < x < \infty.$$

  a Verify that f is a valid probability density function.
  b Sketch the graph of f.
  c Find the mean of X.
  d Find $P(-1 < X < 1)$.

2 A continuous random variable X has a probability density function f given by

$$f(x) = k \cos x, \qquad 0 \leqslant x \leqslant \frac{\pi}{2},$$

$$f(x) = 0, \qquad \text{otherwise.}$$

Find **a** the value of k

b the mean and the variance of X

c $P\left(X > \frac{\pi}{6}\right)$.

3 A continuous random variable X has a probability density function f given by

$$f(x) = \frac{(x-4)}{6}, \qquad 4 \leqslant x \leqslant 6,$$

$$f(x) = \frac{1}{3}, \qquad 6 < x \leqslant 8,$$

$$f(x) = 0, \qquad \text{otherwise.}$$

Find **a** the distribution function of X

b the median

c the lower quartile.

4 A garage has a storage tank of capacity 8000 litres which is filled with petrol each Monday morning. The weekly demand for petrol in thousands of litres is a continuous random variable X with probability density function f given by

$$f(x) = \frac{3(10-x)^2}{1000}, \qquad 0 \leqslant x \leqslant 10,$$

$$f(x) = 0, \qquad \text{otherwise.}$$

a Find the probability that the garage will not be able to meet the demand

(i) next week,

(ii) in exactly one week out of the next four weeks.

b Find the expected number of litres of petrol *sold* per week.

5 P is a point chosen at random on a semi-circle with diameter AB and radius r cm. The size of angle ABP is denoted by X radians and the length of the chord AP is denoted by Y cm. Given that X is uniformly distributed between 0 and $\frac{\pi}{2}$, find

a $P(Y > r)$ **b** $E[Y]$.

10.5 Normal distribution

A continuous random variable X with probability density function f given by

$$f(x) = \frac{1}{\sigma\sqrt{(2\pi)}}\exp\left[-\frac{(x-\mu)^2}{2\sigma^2}\right], \qquad -\infty < x < \infty$$

where $\sigma > 0$, is normally distributed with mean μ and standard deviation σ.

The continuous random variable Z with probability density function ϕ given by

$$\phi(z) = \frac{1}{\sqrt{(2\pi)}}\exp\left[\frac{-z^2}{2}\right], \qquad -\infty < z < \infty$$

is normally distributed with mean 0 and standard deviation 1. Z is called the standard normal variable. The graph of ϕ is symmetrical about $z = 0$ and the area bounded by the graph of ϕ and the z-axis is 1.

Table 3 on page 217 tabulates values of the standard normal distribution function for values of z from 0.00 to 3.49.

Normal approximation to the binomial distribution

If $X \sim B(n, p)$, where n is large and neither p nor $(1 - p)$ is very small, then its distribution may be approximated by the distribution of $Y \sim N(np, np(1 - p))$.

The following approximations are called continuity corrections.

$$P(X = r) \simeq P(r - 0.5 < Y < r + 0.5)$$

$$P(X \leqslant r) \simeq P(Y < r + 0.5)$$

$$P(X \geqslant r) \simeq P(Y > r - 0.5)$$

One rule-of-thumb to determine the conditions under which the use of a normal approximation to a binomial distribution is appropriate is as follows:

$$n \text{ should be greater than the larger of } \frac{16p}{(1 - p)} \text{ and } \frac{16(1 - p)}{p}.$$

Exercise 10.5

1 The times taken by a swimmer to complete n lengths of 50 m are normally distributed with mean $62n$ seconds and standard deviation $2n$ seconds. In a race of 200 m, calculate the probabilities that the swimmer will swim

a the first length in less than 60 seconds

b each of the four lengths in less than 60 seconds (assume the times of the lengths are independent)

c the race in less than 4 minutes.

2 Two filling machines are used to fill 25-kg bags of peat. The first machine produces bags such that the mass of peat in the bags is normally distributed with standard deviation 0.4 kg and the cost of filling each bag is 10p; the second machine produces bags such that the mass of peat in the bags is normally distributed with standard deviation 0.2 kg and the cost of filling each bag is 20p. Both machines are adjusted so that only 1% of the bags produced are underweight. Find which machine is the more economical on average when the price of peat is 10p per kg.

3 Books on the top shelf at a library are directly accessible only to a person having a reachable height of at least 250 cm. It may be assumed that the reachable heights of adult male readers at the library are normally distributed with mean 264 cm and standard deviation 8 cm, and that those of adult female readers are normally distributed with mean 254 cm and standard deviation 5 cm.

(i) Find, correct to three significant figures, the proportion of adult male readers and the proportion of adult female readers who are able to reach books on the top shelf.

(ii) Given that 40 per cent of all adult readers at the library are male, find the proportion, correct to three significant figures, of all adult readers who are able to reach books on the top shelf.

(iii) The library decides to lower the top shelf so that 95 per cent of all adult female readers will be able to reach books there. Find the corresponding percentage, correct to the nearest integer, of all adult male readers who will then be able to reach books on the top shelf.

(*JMB*)

4 In a certain constituency in Manchester, 25% of the electorate voted Liberal in an election. A sample of 200 people are to be interviewed in the constituency and are to be asked how they voted in the election. Let X denote the number in the sample who will say that they voted Liberal. Calculate an approximate value for $P(X > 60)$ and state two assumptions that you have made.

5 An ordinary cubical die, assumed to be fair, is thrown 600 times.

a Find the probability that a six is obtained between 90 and 100 times, inclusive.

b Find the probability that a six is obtained between 90 and 110 times, inclusive.

c Find the two inclusive limits, symmetrical about 100, between which the number of sixes obtained lies with 0.99 probability.

d If 125 sixes were thrown what would you conclude?

Miscellaneous Exercise [10]

1 The following table shows the grouped frequency distribution of the marks obtained by 1000 candidates in an examination.

Marks	0–19	20–29	30–39	40–49	50–59	60–69	70–79	80–100
No. of candidates	68	101	164	239	164	130	82	52

Draw a cumulative frequency diagram to illustrate these data. Showing your method in each case, estimate to the nearest integer
(i) the median mark,
(ii) the pass mark, given that 70 per cent of the candidates pass the examination.

(JMB)

2 The contents of each packet of a certain commodity have a nominal weight of 500 grams. An inspector chose at random a sample of 140 such packets and weighed the contents of each to the nearest gram. The following table shows the grouped frequency distribution of the weights obtained.

Weight	490–494	495–499	500–504	505–509	510–514	515–519
No. of packets	1	3	37	59	34	6

Estimate the probability that a packet chosen at random from the sample contains less than 500 grams.

Draw a histogram to represent the frequency distribution. Prepare a table showing, for the above data, the mid-values of the class intervals and all the terms required for the summations needed to calculate the mean and variance. Calculate the sample mean and variance.

By assuming that the weights are normally distributed, estimate the probability that a packet chosen at random from the total output of packets contains less than 500 grams.

(JMB)

3 A florist plants an equal number of bulbs in each of 80 bowls. After six weeks he counts the number of bulbs which have germinated in each bowl. The results are as follows.

No. of bulbs	0	1	2	3	4	5	6	7	8
No. of bowls	1	3	10	22	25	15	4	0	0

Represent the data graphically.

State formulae for the mean and variance of a frequency distribution, explaining the symbols used. Prepare a table for the given data, showing all the terms in the summations that have to be calculated to obtain the mean and variance, and calculate these quantities for the distribution.

Assuming that each bulb has an equal probability of germinating in the six weeks, state what standard type of distribution you would expect to arise. Assuming that the above data do represent a sample from such a distribution, use the mean and variance you have calculated to estimate
(i) the probability, to three significant figures, of a bulb germinating,
(ii) the number of bulbs planted in each bowl.

(JMB)

4 In a fishing competition, the total catches of 50 anglers had masses (to the nearest 0.1 kg) as given in the following table.

Mass in kg	0–0.2	0.3–0.7	0.8–1.2	1.3–1.7	1.8–2.2	2.3–3.7	3.8–5.2
Frequency	8	8	12	8	8	4	2

Draw a histogram to represent the frequency distribution.

State formulae for the mean and standard deviation of a frequency distribution, explaining the symbols used. Prepare a table for the given data, showing the mid-values of the class intervals and all the terms in the summations that have to be calculated to obtain the mean and variance.

Calculate the mean, variance and standard deviation of the distribution.

Given the additional information that all eight anglers placed in the first class caught nothing at all, obtain a revised value for the mean. *(JMB)*

5 The distribution of the burning times, to the nearest 10 hours, of a sample of 100 electric light bulbs of a particular brand is shown in the following table.

Burning time (to nearest 10 hours)	990	1000	1010	1020	1030	1040
Number of bulbs	5	35	25	20	10	5

(i) Estimate the mean, the variance and the median of the burning times of these 100 bulbs.

(ii) Given a random sample of 5 bulbs of the above brand find an estimate, correct to 3 significant figures, of the probability that at least 4 of them will have burning times of less than 1020 hours. *(JMB)*

6 Three unbiased dice are rolled. Find the probability that the sum of the three scores is 15 or more. Find also the probability that the sum of the three scores is more than 3 and less than 15. *(JMB)*

7 A bag contains five different pairs of gloves. Two persons, *A* and *B*, take turns to draw a glove from the bag (without replacement), *A* drawing first. Find the probabilities

(i) that the first glove drawn by *A* and the first glove drawn by *B* do not form a pair,

(ii) that *A* obtains a pair in his first two draws,

(iii) that *B* obtains a pair in his first two draws,

(iv) that both *A* and *B* obtain a pair in their first two draws,

(v) that at least one of the persons obtains a pair in their first two draws,

(vi) that neither *A* nor *B* obtains a pair in his first two draws,

(vii) that *B* draws a pair on making his second draw and *A* on making his third draw. *(JMB)*

8 An unbiased cubical die has four faces numbered 1 and two faces numbered 2. Two boxes are numbered 1 and 2; the box numbered 1 contains three red discs and two blue discs, and the box numbered 2 contains two red discs and three blue discs. Given that the die is rolled and three discs are then drawn at random without replacement from the box with the same number as that uppermost on the die, calculate the probabilities that

(i) two red discs and one blue disc are drawn from the box numbered 1,

(ii) two red discs and one blue disc are drawn,

(iii) the discs came from the box numbered 1, given that two of the drawn discs were red and one disc was blue. *(JMB)*

9 The two events A and B are independent and their probabilities of occurring are $P(A) = \dfrac{2}{3}$ and $P(B) = \dfrac{2}{5}$.

(i) Show that $P(A \cup B) = \dfrac{4}{5}$.

(ii) Calculate the probability that only one of the two events will occur.

(iii) Calculate the conditional probability that A occurred given that only one of the two events occurred.

Another event C is such that A and C are mutually exclusive, $P(C) = \dfrac{1}{5}$, and $P(B' \cap C') = \dfrac{7}{15}$. Calculate

(iv) $P(B \cap C)$,

(v) $P(B \,|\, A \cup C)$. *(WJEC)*

10 In a biological experiment a large batch of seeds is produced. It is known that the batch contains three types of seed A, B and C in the ratio $6:3:1$. All type A seeds are fertile; one quarter of type B seeds are fertile and the remainder of the seeds of this type are sterile; all type C seeds are sterile.

(i) Find the probability that a seed chosen at random from those produced in the experiment is fertile.

(ii) Given that two randomly selected seeds are both fertile, find the probability that they are both of type B.

(iii) A random sample of n seeds is chosen from the batch so that the probability that the sample contains at least one fertile seed is greater than 0.99. Find the least value of n. *(JMB)*

11 Suppose that 2% of all animals of a certain species have a particular disease. A test can be applied to an animal to determine whether or not it has the disease. When applied to an animal which has the disease, the test will give a positive response (indicating the presence of the disease) with probability 0.96. When applied to an animal which does not have the disease, the test will give a positive response with probability 0.01.

(i) Calculate the probability that when the test is applied to a randomly chosen animal, it will give a positive response.

(ii) Given that a positive response is obtained on an animal calculate, to two decimal places, the probability that the animal has the disease. *(WJEC)*

12 Of the ten cards in a pack, five are numbered 1, three are numbered 2 and the remaining two cards are numbered 3. Three cards are selected at random without replacement from the pack. Calculate the probabilities that

   (i) all three selected cards have the same number,

   (ii) the numbers on the three selected cards are 1, 2 and 3 (in any order),

   (iii) the largest (or equal largest) of the numbers on the selected cards is 2.

   Let X denote the sum of the numbers on the three selected cards.

   (iv) Find the probability distribution of X.

   (v) Deduce the most probable value of the sum of the numbers on the seven cards remaining in the pack. *(WJEC)*

13 In a multiple-choice test paper consisting of 20 questions, an examinee has to choose which one of five listed answers to each question is correct. Each correct answer earns 4 marks and 1 mark is deducted for each incorrect answer.

   a One examinee chooses an answer to each question at random from those listed. Find the probability, to three decimal places, that this examinee's total mark on the paper will be 10 or more.

   b Another examinee is able to identify three of the listed answers in each of n of the questions as being incorrect, and for each of these questions she chooses one of the remaining two answers at random. For each of the remaining $(20 - n)$ questions she is unable to identify any of the listed answers as being incorrect and chooses her answer at random from the five listed. Find the smallest value of n for which the mean total mark that this examinee can obtain is 20 or more. *(JMB)*

14 The probability distribution of a discrete random variable X is given by

$$P(X = r) = \frac{1}{n}, \qquad r = 1, 2, 3, \ldots, n.$$

   Find, in terms of n, the mean of X, and show that the variance of X is $\frac{1}{12}(n^2 - 1)$. *(JMB)*

15 A farmer has a flock of N sheep of which 4 are black and the others white. In a random sample of 3 sheep chosen without replacement, let X denote the number that are black.

   (i) Find the probability distribution of X.

   (ii) Show that the expected value of X is $\frac{12}{N}$.

   (iii) Determine the set of values of N for which $X = 1$ is more probable than any of the other possible values of X. *(JMB)*

16 At a side-stall in a fete, a penny of diameter 2 cm is thrown on to a square board of side 60 cm, the board having raised edges so as to ensure that the penny will not go off the board. The board is divided into 400 identical squares each of side 2.6 cm; squares adjacent to the edge of the board are 2 mm from the edge and adjacent squares are 4 mm apart. The diagram, which is not to scale, shows the layout at one corner of the board.

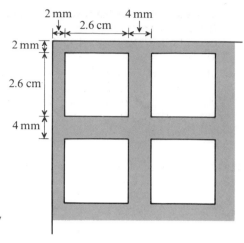

Assume that the centre of a penny thrown on to the board is equally likely to come to rest at any possible location on the board.

(i) Show that the probability of a penny coming to rest completely within a square is $\dfrac{36}{841}$.

(ii) Of the 400 squares, 200 are marked '2p', 100 are marked '3p', 60 are marked '5p', and the remaining 40 are marked '10p'. The stall-holder keeps every penny that is thrown on to the board, but when a penny comes to rest completely within a square the stall-holder pays out the amount marked in that square. Calculate the stall-holder's expected gain in 100 throws, giving your answer correct to the nearest penny.

(*JMB*)

17 A particular species of bird lays from one to five eggs. The probability of r eggs is proportional to r. A collector of eggs takes one egg from a nest if there are more than three eggs there and none otherwise. Show that the probability that he takes an egg is $\dfrac{3}{5}$.

In an area he finds six occupied nests of this species which have not previously been interfered with. Find, in fractional form, the probability that he obtains a total of m eggs from the six nests for $m = 0, 1, 2, 3, 4, 5, 6$, respectively.

Obtain the mean of this distribution and determine the probabilities that he gets
(i) fewer than four eggs,   (ii) more than four eggs.

A second collector, obeying the same rule for collecting eggs, follows the first collector round the same six nests. Find the mean number of eggs the second collector would expect to collect. (*JMB*)

18 Three independent events A, B, C are such that $P(A) = 0.5$, $P(B) = 0.25$ and $P(C) = 0.2$. Show that $P(A \cup B \cup C) = 0.7$.

A quiz team has three members, Ann, Beryl and Charles. In response to each question, each member of the team writes down an answer without

consultation. The team scores one point whenever at least one of their three answers to a particular question is correct. Independently, the probabilities that Ann, Beryl and Charles will answer a question correctly are 0.5, 0.25 and 0.2 respectively.
  (i) Use tables to find, to three decimal places, the probability that the team will score at least 12 points in a quiz consisting of 20 questions.
  (ii) Use a suitable approximate method to find, to three decimal places, the probability that the team will score at least 60 points in a quiz consisting of 100 questions. *(JMB)*

19 A tennis team takes part in a competition in which it plays four matches at home and four matches away. Independently for each match played the probability that the team will win a home match is 0.8 and the probability that it will win an away match is 0.3.
  (i) Calculate the probability that the team's first home win will be the third home match it plays.
  (ii) Calculate, to four decimal places, the probability that in the competition the team will end up having won exactly two of its home matches and exactly two of its away matches.
  (iii) Calculate, to four decimal places, the probability that in the competition the team will win 3 more home matches than away matches.
  (iv) Let T denote the total number of matches (home and away) that the team will win in the competition. Find the mean and the variance of T. Hence, or otherwise, show that T is not binomially distributed. *(WJEC)*

20 The continuous random variable X has the probability density function f, where
$$f(x) = \frac{1}{2}(x - 2), \qquad \text{for } 2 \leqslant x \leqslant 4,$$
$$f(x) = 0, \qquad \text{otherwise.}$$
By first expanding $(X - c)^2$, or otherwise, find the two values of c such that $E[(X - c)^2] = \frac{2}{3}$. *(WJEC)*

21 The probability density function f of a continuous random variable X is given by
$$f(x) = k \sin x, \qquad 0 \leqslant x \leqslant \pi,$$
$$f(x) = 0, \qquad \text{otherwise.}$$
  (i) Show that $k = \frac{1}{2}$.
  (ii) Find the mean of X.
  (iii) Show that the variance of X is $\frac{1}{4}\pi^2 - 2$.
  (iv) Calculate, to three decimal places, the difference between the 10th percentile of X and the 90th percentile of X. *(JMB)*

195

22 A retailer finds that the weekly demand, X kg, by his customers for a perishable commodity has the probability density function

$$f(x) = kx^2 e^{-ax}, \qquad (0 \leqslant x < \infty), \quad a > 0$$

$$f(x) = 0, \qquad (x < 0).$$

Show that $k = \frac{1}{2}a^3$ and that the mean of the distribution is $\sqrt{3}$ times the standard deviation of the distribution.

$$\left[\text{You may use the result } \int_0^\infty x^r e^{-ax} \, dx = \frac{r!}{a^{r+1}}, \text{where } a > 0 \text{ and } r = 0,1,2,\ldots \right].$$

Sketch the graph of the probability density function.

Given that the mean weekly demand is 60 kg, deduce the value of a and find the probability that the retailer has some goods left a week after delivery if he buys 40 kg of the commodity.

(JMB)

23 A variable X has a probability density function given by

$$0 \quad \text{for } x < \alpha$$

$$f(x) \quad \text{for } \alpha \leqslant x \leqslant \beta$$

$$0 \quad \text{for } x > \beta$$

(x represents an actual value taken by the random variable X.)

Give two conditions that must be satisfied by the function $f(x)$.

In the case where $\alpha = 0$, $\beta = 1$ and $f(x) = kx(x - 1)^2$, determine the value of k. Show that there is no suitable value for k when $\alpha = -1$ and $\beta = 1$.

Calculate the mean and variance of X for the distribution for which k has been determined. Find the value of X for which the probability density function is a maximum and sketch the probability density function.

(JMB)

24 The mass X kg of a particular substance produced per hour in a chemical process is a continuous random variable whose probability density function is given by

$$f(x) = \frac{3x^2}{32}, \qquad 0 \leqslant x < 2,$$

$$f(x) = \frac{3(6 - x)}{32}, \qquad 2 \leqslant x \leqslant 6,$$

$$f(x) = 0, \qquad \text{otherwise.}$$

(i) Find the mean mass produced per hour.

(ii) The substance produced is sold at £2 per kg and the total running cost of the process is £1 per hour. Find the expected profit per hour and the probability that in an hour the profit will exceed £7.

(JMB)

25 The distribution of the salaries, in thousands of pounds, of persons in a certain profession can be approximated by a continuous distribution having probability density function f given by

$$f(x) = \frac{1}{90}x(9 - x), \qquad 3 \leqslant x \leqslant 9,$$

$$f(x) = 0, \qquad\qquad\qquad \text{otherwise.}$$

(i) Find the mean salary and show that more than half the persons in the profession earn less than the mean salary.

(ii) Each person in the profession has to contribute 5 per cent of his/her salary towards a pension scheme. Find the mean amount contributed. It is decided to raise this mean contribution to £305, by requiring each person earning more than £6000 to pay a fixed amount £C in addition to the 5 per cent of his/her salary. Find the value of C.

(JMB)

26 The operational lifetimes in hundreds of hours of a battery-operated minicalculator may be regarded as a continuous random variable having probability density function

$$f(x) = cx(10 - x), \qquad 5 \leqslant x \leqslant 10,$$

$$f(x) = 0, \qquad\qquad\qquad \text{otherwise.}$$

a Find the value of c and of the expected operational lifetime of such a minicalculator.

b The purchase price of such a minicalculator is £20 and its running cost (for batteries) amounts to 20 pence per hundred hours of operation. Thus, the overall average cost in pence per hundred hours of operation of a minicalculator whose operational lifetime is X hours is given by

$$Y = 20 + \left(\frac{2000}{X}\right).$$

(i) Evaluate E(Y), the expected overall average cost per hundred hours.

(ii) Find the probability that the overall average cost per hundred hours will exceed £2·70.

(WJEC)

27 The time, in minutes, required to complete a particular task may be assumed to be normally distributed. Given that there is a probability of 0.01 that the task will be completed in less than 30 minutes and a probability of 0.95 that it will be completed in less than one hour, calculate, correct to two decimal places, the probability that the task will be completed in less than 45 minutes.

(JMB)

28 The number X of air bubbles in a mass-produced lens is a discrete random variable such that

$$P(X = r) = \frac{1}{3} - \frac{r}{15}, \qquad r = 0, 1, 2, 3, 4.$$

A lens which has two or more air bubbles is rejected as unsuitable. Evaluate the probabilities that

(i) a randomly chosen lens will be rejected,

(ii) a lens chosen at random from lenses that have been rejected will have exactly two air bubbles.

A lens having one or no air bubble is then tested to determine its dispersion index. Such a lens is passed only if it has *either* no air bubble and a dispersion index less than 4 *or* one air bubble and a dispersion index less than 3.5. It may be assumed that, independently of the number of air bubbles, the dispersion indices of the lenses are normally distributed with mean 3.6 and standard deviation 0.5.

(iii) Find the proportion of all lenses that are passed, giving your answer correct to three decimal places.

(iv) A lens is chosen at random from those that were passed. Find the probability that it has no air bubble, giving your answer correct to three decimal places.

(v) A random sample of twenty lenses is chosen from those lenses that are passed. Find, correct to two decimal places, the mean and the variance of the proportion of the lenses in the sample that have no air bubble. *(JMB)*

29 The operational lifetimes, in hours, of certain electronic components are normally distributed with mean μ and standard deviation 200.

a Find the value of μ if 5 per cent of the components will operate for less than 1200 hours.

b Suppose that $\mu = 1512$ and that a good component is one having a lifetime in excess of 1600 hours. Find the proportion of components that are good.

If components are sold in batches of 100,

(i) name the distribution of the number of good components per batch, and find the mean and the variance of the proportion of good components per batch;

(ii) find an approximate value for the probability that a batch will contain at least 40 good components. *(JMB)*

30 The breaking strengths of pieces of light inextensible string produced by a new manufacturing process have a normal distribution with mean 54 N and standard deviation 10 N. A 4-kg mass is suspended from a single piece of this string. Calculate (to 3 decimal places) the probability that the string breaks, taking g as $9.8\,\mathrm{ms}^{-2}$.

In the case when the string is not broken, the free end is passed over a smooth pulley and attached to a 10-kg mass. Initially the masses are held so that the parts of the string not in contact with the pulley are just taut and vertical. Calculate (to 3 decimal places) the probability that the string breaks when the masses are released.

31 The random variable X has the probability density function f defined by

$$f(x) = xe^{-x}, \qquad x \geqslant 0,$$
$$f(x) = 0, \qquad x < 0.$$

Show that, for $x \geqslant 0$, the distribution function F is given by

$$F(x) = 1 - (1 + x)e^{-x}.$$

Denoting the 25th percentile of X by t, show that

$$3 - 4(1 + t)e^{-t} = 0.$$

Using Newton's method and taking 1 as a first approximation, find the root of this equation correct to two decimal places.

(*JMB*)

32 The random variable X has the probability density function defined by

$$f(x) = k(x^2 - 3), \qquad 1 \leqslant x \leqslant 2,$$
$$f(x) = 0, \qquad \text{otherwise.}$$

Show that the median m of X satisfies the equation

$$x^3 - x - 3 = 0.$$

Show that this equation has only one real root and verify that this root lies between 1 and 2.

Use Newton's method with a starting point of 1.5 and two iterations to find (to 2 decimal places) an approximate value for the median m.

Answers

The Examining Boards listed in the Acknowledgements on page ii bear no responsibility whatever for the answers to examination questions given here, which are the sole responsibility of the author.

Chapter 1

Descriptive statistics

Exercise 1.1

1

No. of goals	0	1	2	3	4	5	6	7
Frequency	4	8	8	8	7	4	4	3

2

No. of children	0	1	2	3	4	5
Frequency	15	13	10	8	3	1

Exercise 1.2

1

No. of goals	0	1	2	3	4	5	6	7
Cum. freq.	4	12	20	28	35	39	43	46

2

No. of children	0	1	2	3	4	5
Cum. freq.	15	28	38	46	49	50

3

U.C.B.	5	11	18	30	40	60	80
Cum. freq.	428	572	733	973	1257	1438	1500

4

U.C.B.	74.5	79.5	84.5	89.5	94.5	99.5	109.5
Cum. rel. freq.	0.16	0.37	0.65	0.84	0.92	0.98	1.00

5

U.C.B.	29.5	44.5	59.5	74.5	89.5	119.5
Cum. freq.	56	119	206	329	396	418

6

U.C.B.	0.965	0.975	0.985	0.995	1.005	1.015	1.025	1.035
Cum. freq.	5	17	34	55	79	99	113	120

7

U.C.B	5.5	9.5	13.5	17.5	21.5	25.5	35.5
Cum. freq.	9	22	41	56	66	74	76

8

U.C.B.	12	13	14	15	16	17	18	19
Cum. freq.	182	357	547	728	896	979	1043	1050

Exercise 1.3

1 Qualitative
2 Ratio, continuous
3 Ratio, discrete
4 Ordinal
5 Ratio, continuous

6 Interval, discrete
7 Ordinal
8 Interval, discrete
9 Qualitative
10 Ratio, continuous

Miscellaneous Exercise 1

1
Score	1	2	3	4	5	6
Freq.	10	10	14	6	9	11
Cum. freq.	10	20	34	40	49	60

2
U.C.B.	114.5	119.5	124.5	129.5	134.5	139.5	144.5	154.5
Cum. freq.	10	23	39	63	90	111	126	141

3
No. of particles	0	1	2	3	4	5	6	7	8	9	10
Cum. freq.	2	9	26	55	91	126	156	177	190	197	200

4
U.C.B.	49.5	54.5	59.5	64.5	69.5	79.5	89.5
Cum. freq.	4	9	19	31	36	42	50

5
Class	0–9	10–19	20–29	30–39	40–49	50–59	60–69	70–79	80–89	90–99
Freq.	4	4	6	12	19	23	12	9	7	4
Cum. freq.	4	8	14	26	45	68	80	89	96	100

Chapter 2

Summary measures

Exercise 2.1

1 3.27, 1.99
2 48.40, 1.82
3 51.17, 12.14
4 130.5, 10.30
5 3.333, 0.0366

6 0.9965, 0.0183
7 13.36, 6.72
8 13.99, 1.81
9 41.74, 24.44
10 25.396, 0.0157

Exercise 2.2

1 6, 7
2 16.5, 14
3 47, 14
4 19.5, 9
5 5.5, 6

6 3, 3
7 49, 3
8 51.6, 16.5
9 130.8, 14.6
10 3.334, 0.048

Miscellaneous Exercise 2

1 a 8, 8.8   **b** (i) 9, 8.8   (ii) 16, 35.2   (iii) 23, 79.2

2 a 2, 0.6   **b** (i) 4, 0.6   (ii) 6, 5.4   (iii) 5, 2.4

3 35, 18.5;   40, 18.5 years   **5** 83, 4.96 kg   **7** 44, 21

4 44.6, 18.87   **6** 8, 7.28;   5, 4.47 years

8
U.C.B.	7	11	17	19	
Cum. freq.	613	1840	3404	3500	10.7, 6.16 years

9
U.C.B.	45	46	47	48	49	50	51	52	
Cum. freq.	24	55	93	140	215	264	290	300	49, 3

10 671, 357   **11** 2.694, 1.422

12 U.C.B.   19.5   39.5   59.5   79.5   100

Cum. freq.   31   87   176   239   263   50, 36

13 U.C.B.   9.5   19.5   29.5   39.5   50

Cum. freq.   7   25   60   85   100   27, 16, 23

14 b 3.86, 3.65   **c** 4, 3

15 Median 63,

M.I.V.   32.5 37.5 42.5 47.5 52.5 57.5 62.5 67.5 72.5 77.5 82.5 87.5 92.5

Freq.   1   3   7   14   22   32   35   32   25   16   8   3   2

mean 63.05 kg, standard deviation 11.28 kg

16 b £5970   **c** £5527, not distorted by atypical values, £1606

17 a 18.6, 4.5   **b** 13.3   **c** 13.925 years

18 a 19 years 7.5 months (N.B. Recorded ages are on average 6 months less than actual age)

b (i) 22 years 1 month   (ii) 11 months

19 U.C.B.   76.5   79.5   82.5   85.5   88.5   91.5   94.5

Cum. freq.   4   11   36   57   69   77   80

(i) median = 83 min   (ii) 6 min   (iii) 24%;   9.00 a.m.

20 6.49, 1.714, 7

21 U.C.B.   13m 29.5s   13m 59.5s   14m 29.5s   14m 59.5s   15m 29.5s

Cum. freq.   2   5   21   44   69

U.C.B.   15m 59.5s   16m 29.5s   16m 59.5s   17m 29.5s

Cum. freq.   87   94   98   100

15 min 08.5 s, 2304 s^2, 48 s;   15 min 07 s

22 5.5 min, 3.83 min

23 U.C.B.   19.5   29.5   39.5   49.5   59.5   69.5   79.5   89.5

Cum. freq.   4   22   50   106   131   146   149   150

44.0 kg, 13.3 kg;   44 kg;   3rd and 97th

24 11 years 11 months, 5 years 9 months

25 57 min, 71.5 min, 32%

26 5.97, 18.12

Chapter 3

Probability

Exercise 3.1

1 a $\dfrac{3}{8}$ **b** $\dfrac{1}{8}$ **4 a** $\dfrac{3}{8}$ **b** $\dfrac{3}{16}$

2 a $\dfrac{1}{5}$ **b** $\dfrac{1}{2}$ **c** $\dfrac{3}{5}$ **5 a** $\dfrac{1}{6}$ **b** $\dfrac{11}{12}$ **c** 0

3 a $\dfrac{1}{2}$ **b** $\dfrac{1}{4}$

Exercise 3.2

1 **a** 0.3 **b** 0.8

2 **a** 0.9 **b** 0.1

3 **a** 0.2 **b** 0.1

4 **a** 0.8 **b** 0.3

Exercise 3.3

1 **a** 0.2 **b** 0.5

2 **a** 0.2 **b** 0.7

3 **a** 0.06 **b** $\dfrac{3}{13}$

5 **a** 0.04 **b** 0.76

Miscellaneous Exercise 3

1 (i) 0.02 (ii) 0.78 (iii) 0.76 (iv) $\dfrac{1}{30}$

2 (i) 0.3 (ii) 0.3 (iii) 0.5

3 0.128, 0.572, 0.2867, 0.4267, 0.428, 0.872

4 (i) 0.02 (ii) 0.45 (iii) 0.6

5 (i) $\dfrac{1}{12}$ (ii) $\dfrac{3}{4}$ (iii) $\dfrac{1}{4}$

6 (i) 0.2 (ii) 0.1; $\dfrac{1}{6}$

7 (i) $\dfrac{2}{5}$ (ii) $\dfrac{3}{10}$ (iii) $\dfrac{3}{10}$; $\dfrac{7}{25}$

8 (i) p^2 (ii) $\dfrac{(1-p)^2}{25}$ (iii) $\dfrac{2}{5}(2+3p)(1-p)$

9 (i) $P(A \cap B) = P(A) \cdot P(B)$ (ii) $P(A \cap B) = 0$ or $P(A \cup B) = P(A) + P(B)$
0.52, 0.5, $\dfrac{9}{35}$, 0.3

10 $\dfrac{1}{6}$

11 (i) not independent (ii) not mutually exclusive, $\dfrac{671}{1296}$

12 **a** $\dfrac{7}{12}$ **b** (i) $\dfrac{1}{4}$ (ii) $\dfrac{6}{7}$

13 $\dfrac{1}{2}, \dfrac{4}{9}, \dfrac{1}{6}$

14 (i) $\dfrac{2}{3}$ (ii) $\dfrac{2}{3}$ (iii) $\dfrac{1}{3}$

15 (i) $\dfrac{5}{12}$ (ii) $\dfrac{5}{12}$ (iii) $\dfrac{1}{6}$

16 (i) $\dfrac{9}{38}$ (ii) $\dfrac{10}{19}$ (iii) $\dfrac{1}{19}$

17 (i) $\dfrac{1}{2}$ (ii) $\dfrac{1}{3}$ (iii) $\dfrac{2}{3}$ (iv) $\dfrac{4}{7}$

18 **a** (i) $P(A) + P(B)$ (ii) $P(A) + P(B) - P(A) \cdot P(B)$
b (i) 0.68 (ii) 0.33; 0.16

19 (i) 0.35 (ii) $\dfrac{4}{13}$

Chapter 4

Permutations and combinations

Exercise 4.1

1 120
2 6.04152×10^{52}
3 6720
4 1.27511×10^{49}
5 526748400
6 40
7 870
8 900
9 39916800
10 (i) 64 (ii) 24
11 (i) 48 (ii) 18
12 24
13 50400
14 9979200
15 (i) 576 (ii) 40320

Exercise 4.2

1 455
2 4.21171×10^{11}
3 2225895
4 a 120 b 120
5 15504
6 190
7 90
8 680
9 112
10 1120
11 120
12 126
13 a 60 b 10
14 a 125 b 35

Miscellaneous Exercise 4

1 44100
2 30240
3 14175
4 111
5 531441

6 (i) 15500 (ii) 480
7 234
8 78
9 159; 640
10 3300; 2730

Chapter 5

More probability

Miscellaneous Exercise 5

1 (i) $\frac{5}{12}$ (ii) $\frac{8}{15}$ (iii) $\frac{1}{12}$

2 (i) $\frac{65}{81}$ (ii) $\frac{11}{27}$ (iii) $\frac{25}{216}$ (iv) $\frac{1}{54}$ (v) $\frac{109}{216}$

3 a (i) $\frac{1}{6}$ (ii) $\frac{5}{18}$ (iii) $\frac{1}{6}$ b $\frac{6}{35}$ c $\frac{23}{42}$

4 (i) $\frac{9}{10}$ (ii) $\frac{1}{5}$ (iii) $\frac{23}{35}$ (iv) $\frac{8}{35}$ (v) $\frac{3}{16}$

5 (i) $\frac{8}{27}$ (ii) $\frac{1}{9}$ (iii) $\frac{53}{54}$ (iv) $\frac{37}{216}$

6 (i) $\dfrac{5}{192}$ (ii) $\dfrac{5}{324}$ (iii) $\dfrac{63}{64}$ (iv) $\dfrac{5}{216}$

7 $\dfrac{74}{81}$

8 (i) 0.4   (ii) 0.4752   (iii) 6

9 a (i) $\dfrac{1}{7}$ (ii) $\dfrac{6}{7}$ (iii) $\dfrac{1}{35}$ **b** $\dfrac{1}{30}, \dfrac{1}{12}$

10 a (i) 0.52   (ii) 0.49   **b** (i) 0.0039   (ii) 1.65×10^{-8}

11 5   **12** $\dfrac{(p-q)}{p}, \dfrac{p}{(p+q)}$

13 (i) $\dfrac{2}{261}$ (ii) $\dfrac{16}{609}$; $\dfrac{308}{435}, \dfrac{204}{1015}$

14 (i) $\dfrac{1}{4}$ (ii) $\dfrac{1}{28}$ (iii) $\dfrac{1}{4}$; $\dfrac{11}{36}$;   (iv) $\dfrac{11}{144}$ (v) $\dfrac{1}{28}$ (vi) $\dfrac{3}{7}$

15 (i) $\dfrac{1}{5}$ (ii) $\dfrac{1}{3}$; $\dfrac{25}{63}$

16 (i) $\dfrac{3}{8}$ (ii) $\dfrac{1}{2}$ (iii) $\dfrac{1}{6}$; $\dfrac{7}{50}$

17 (i) $\dfrac{31}{120}$ (ii) $\dfrac{91}{240}$ (iii) 1; $\dfrac{2}{7}$

18 a (i) $\dfrac{3}{11}$ (ii) $\dfrac{3}{7}$ **b** (i) $\dfrac{2}{35}$ (ii) $\dfrac{1}{14}$

19 (i) $\dfrac{5}{9}$ (ii) $\dfrac{3}{4}$

20 (i) $\dfrac{1}{14}$ (ii) $\dfrac{3}{7}$ (iii) $\dfrac{1}{30}$

21 0.624   **22** 1

23 a (i) $\dfrac{1}{21}$ (ii) $\dfrac{41}{42}$ (iii) $\dfrac{1}{30}$ **b** $\dfrac{1}{31}$

24 a 0.5, 0.3   **b** 0.3

25 (i) $\dfrac{2}{15}$ (ii) $\dfrac{1}{2}$

26 a (ii) $\dfrac{3}{4}, \dfrac{3}{10}$

27 0.59   (i) 0.352   (ii) 0.4576   (iii) 0.48064   **28** 0.6336, 0.5508

29 (i) $\dfrac{1}{11}$ (ii) $\dfrac{21}{55}$ (iii) $\dfrac{3}{11}$ (iv) $\dfrac{2}{55}$ (v) $\dfrac{36}{55}$ (vi) $\dfrac{5}{11}$

30 (i) $\dfrac{7}{24}$ (ii) $\dfrac{1}{40}$ (iii) $\dfrac{1}{120}$

31 b (i) 0.375   (ii) neither

32 (ii) $\frac{1}{3}$ (iii) $\frac{17}{64}$

33 (i) $\frac{14}{55}$ (ii) $\frac{9}{55}$ (iii) $\frac{6}{55}$; $\frac{35}{99}$; $\frac{9}{55}$

34 a (i) $\frac{5}{18}$ (ii) $\frac{1}{6}$ (iii) $\frac{5}{14}$ (iv) $\frac{71}{126}$ **b** 18

35 $\frac{11}{15}$ (i) $\frac{47}{57}$ (ii) $\frac{1}{3}$

36 (i) $\frac{5}{42}$ (ii) $\frac{2}{7}$ (iii) $\frac{5}{84}$

37 $\frac{(n-1)}{2(2n-1)}$

38 (i) $2^{r-1}\frac{(N-1)!}{(N-r)!}\frac{(2N-r)!}{(2N-1)!}\frac{r}{(2N-r)}$; (ii) $\frac{3}{(2N-1)(2N-3)}$

39 a (i) $\left(\frac{N-1}{N}\right)^{k-1}$ (ii) $\frac{N!}{(N-k)!\,N^k}$

Chapter 6

Discrete random variables

Exercise 6.1

x	0	1	2	3
1 p	$\frac{1}{56}$	$\frac{15}{56}$	$\frac{30}{56}$	$\frac{10}{56}$

x	1	2	3	4	5	6
3 p	$\frac{11}{36}$	$\frac{9}{36}$	$\frac{7}{36}$	$\frac{5}{36}$	$\frac{3}{36}$	$\frac{1}{36}$

x	0	2	4
2 p	$\frac{3}{8}$	$\frac{4}{8}$	$\frac{1}{8}$

x	1	2	3	4
4 p	$\frac{35}{56}$	$\frac{15}{56}$	$\frac{5}{56}$	$\frac{1}{56}$

5 $P(X = x) = \left(\frac{1}{4}\right)\left(\frac{3}{4}\right)^{x-1}$ $x = 1, 2, 3, \ldots$

Exercise 6.2

1 $\frac{15}{8}$ **4** $\frac{3}{2}$

2 $\frac{3}{2}$ **5** 4

3 $\frac{91}{36}$

Exercise 6.3

1 1.6, 1.44

2 5, 4

3 $\frac{17}{7}, \frac{40}{49}$

4 25, 125

5 3, 2.4

Exercise 6.4

1 **a** 6, 4   **b** 6, 16   **c** 9, 16   **d** 7, 36   **e** -3, 16   **f** 0, 1

2 **a** 6, 2   **b** 15, 6   **c** 16, 6   **d** -1, 2   **e** 17, 8   **f** 0, 1

3 $\dfrac{6}{5}, \dfrac{32}{75}; \dfrac{26}{5}, \dfrac{32}{75}$

4 £70, £6

5 0, 400

Exercise 6.5

1 **a** 9   **b** 1   **c** 14   **d** 31

2 **a** 1   **b** 5   **c** 4   **d** 1

3 **a** 12   **b** 5   **c** 4

4 £130

5 180

Miscellaneous Exercise 6

1 2, 2

2 $\dfrac{17}{7}, \dfrac{26}{49}$

3 $\dfrac{1}{5}, \dfrac{3}{10}$

4 $\dfrac{125}{216}, \dfrac{75}{216}, \dfrac{15}{216}, \dfrac{1}{216};$ loss 2p

5 $\dfrac{391}{1080}$

6 5.25, 3.22; 1

7 (i) 0.363   (ii) 0.883;   1.6

9

x	0	1	2	3	4	5
p	$\dfrac{243}{1024}$	$\dfrac{405}{1024}$	$\dfrac{270}{1024}$	$\dfrac{90}{1024}$	$\dfrac{15}{1024}$	$\dfrac{1}{1024}$

1.25, 75p

11 **a** 8   **b** 2p loss

12 $\dfrac{23}{35}, \dfrac{2}{3}, 1.8$

13 $\dfrac{(n-1)}{3},$ (i) $\dfrac{37}{105}$ (ii) $\dfrac{4}{7}$

14 $\dfrac{1}{4}, \dfrac{15}{4}$

(i)

x	2	3	4	5	6	7	8	9	10	11	12
p	$\dfrac{1}{72}$	$\dfrac{4}{72}$	$\dfrac{5}{72}$	$\dfrac{8}{72}$	$\dfrac{9}{72}$	$\dfrac{12}{72}$	$\dfrac{11}{72}$	$\dfrac{8}{72}$	$\dfrac{7}{72}$	$\dfrac{4}{72}$	$\dfrac{3}{72}$

(ii) $\dfrac{3}{11}$ (iii) $\dfrac{67}{72}$

15 (i) 0.000048   (ii) 0.00018624; $P(X = r) = 0.02 \times (0.98)^{r-1}$, 0.18

16 (i) $\dfrac{1}{21}$ (ii) $\dfrac{2}{7}$ (iii) $\dfrac{4}{9}$ (iv) $\dfrac{5}{42}$ (v) 1   (vi) $\dfrac{4}{3};$ 0

17 **a** (i) $\dfrac{21}{38}$ (ii) $\dfrac{5}{19}$ (iii) $\dfrac{7}{38}$ **b** $\dfrac{265}{38}$

18 (ii) $\dfrac{1}{16}, \dfrac{3}{32}, \dfrac{3}{64}$ (iii)

x	2	3	4	
p	$\dfrac{6}{16}$	$\dfrac{7}{16}$	$\dfrac{3}{16}$	$3, \dfrac{45}{16}$

19 $\dfrac{2(n+1)}{3}$

20 (i) p^2 (ii) 0   (iii) $2p^3(1-p)$; $\dfrac{2}{(1-2p+2p^2)}$

21 (i) $\dfrac{1}{25}$ (ii) $\dfrac{9}{100}$ (iii) $\dfrac{3}{20}$;   (i) 0.0115   (ii) 0.0097   (iii) 0.0840

22 $\dfrac{1}{4}$; $-\dfrac{1}{3}, \dfrac{43}{18}$

Chapter 7

Special discrete distributions

Exercise 7.1

1 a 0.632   **b** 0.238   **6 a** 0.777   **b** 0.037

2 a 0.956   **b** 0.172   **7 a** 0.900   **b** 0.234

3 a 0.238   **b** 0.028   **8 a** 0.905   **b** 0.665

4 a 0.867   **b** 0.285   **9 a** 0.977   **b** 0.774

5 a 0.653   **b** 0.289   **10 a** 0.104   **b** 0

Exercise 7.2

1 6

2 22.4, 23

3

x	0	1	2	3
p	4	12	12	4

4

x	0	1	2	3	4
p	625	500	150	20	1

5 $B(100, 0.25)$, 25, 18.75; 110, 75

Exercise 7.3

1 $\dfrac{n}{2}, \dfrac{n(n+2)}{12}$ **4** $\dfrac{9}{19}, \dfrac{6}{19}, \dfrac{4}{19}$; 3, 6; $\dfrac{7}{15}, \dfrac{5}{15}, \dfrac{3}{15}$

2 0.42, 0.0247

5 $P(X = r) = n\left(\dfrac{1}{2}\right)^{n-1}\left[1 - n\left(\dfrac{1}{2}\right)^{n-1}\right]^{r-1}$

3 $\dfrac{5}{3}, \dfrac{5}{9}$; $\dfrac{4}{3}, \dfrac{5}{9}$

Miscellaneous Exercise ☐ 7

1 4

2 6

3 (i) 0.349   (ii) 0.387   (iii) 0.194   (iv) 0.070

4 (i) 0.001   (ii) 0.015   (iii) 0.088   (iv) 0.263   (v) 0.396   (vi) 0.237;   0, 1, 9, 26, 40, 24

5 0.15;   44, 39, 14, 3, 0, 0

6 (i) 0.264   (ii) 0.690

7 0.882

8 9.95, 1.8275;   0.829;   B(12, 0.829)

9 19.5, 1.015;   0.975;   B(20, 0.975)

10 0.107, 0.269, 0.302, 0.201, 0.088, 0.027, 0.005, 0.001, 0, 0, 0;   2;   0;   0.4, 4, 10

11 $(1 - p)^{10}$;   0.865

12 0.0016, 0.0256, 0.1536, 0.4096, 0.4096;   (i) 0.973   (ii) 1.000; $0, 0, \dfrac{2}{15}, \dfrac{8}{15}, \dfrac{5}{15}$

13 (i) 0.02   (ii) 0.77   (iii) 0.029   (iv) 0.05;   £1500;   0.228

14 1.5, 0.75;   0.05, 0.45, 0.45, 0.05;   60%

15 $\dfrac{1}{64}, \dfrac{6}{64}, \dfrac{15}{64}, \dfrac{20}{64}, \dfrac{15}{64}, \dfrac{6}{64}, \dfrac{1}{64}$;   (i) $\dfrac{231}{1024}$ (ii) $\dfrac{4017}{4096}$

16 (i) 0.1296   (ii) 0.1792;   0.25, 0.50, 0.25;   0.4816

17 $\dfrac{1}{12}, \dfrac{1}{72}$;   0.59049, 0.32805, 0.07290, 0.00810, 0.00045, 0.00001;

    (i) 0.00046   (ii) 0.99144

18 (i) $\dfrac{64}{125}, \dfrac{48}{125}, \dfrac{12}{125}, \dfrac{1}{125}$;   0.6   (ii) $\dfrac{220}{455}, \dfrac{198}{455}, \dfrac{36}{455}, \dfrac{1}{455}$;   0.6

19 (i) B(n, p), np, np(1 − p + np)   (ii) $\dbinom{n}{r}(1 - p)^{n-r} p^r \dfrac{r}{n}$ (iii) $\dbinom{n}{r}(1 - p)^{n-r} p^r \dfrac{r}{np}$

20 **b** (i) $\dfrac{1}{14}$ (ii) £4·80   **c** 25, £5·00

21 (i) $(1 - 2p)^n$ (ii) $1 - (1 - 2p)^n$

Chapter 8

Continuous random variables

Exercise ☐ 8.1

1 **a** $\dfrac{1}{2}$ **b** $\dfrac{3}{4}$ **c** $\dfrac{9}{16}$ **4** **a** $\dfrac{1}{4}$ **b** $\dfrac{15}{16}$ **c** 0

2 **a** $\dfrac{1}{9}$ **b** $\dfrac{16}{27}$ **c** $\dfrac{11}{27}$ **5** f(x) < 0   for 5 < x ⩽ 6

3 **a** $\dfrac{1}{4}$ **b** $\dfrac{1}{2}$ **c** $\dfrac{1}{2}$

Exercise 8.2

1 $\dfrac{2}{3}, \dfrac{2}{9}$

2 $\dfrac{7}{4}, \dfrac{43}{80}$

3 $6, \dfrac{4}{3}$

4 $\dfrac{8}{5}, \dfrac{8}{75}$

5 $\dfrac{51}{28}, \dfrac{1891}{3920}$

Exercise 8.3

1 $F(x) = \dfrac{x^2}{4}, \quad 0 \leqslant x \leqslant 2; \quad \sqrt{2}; \quad \sqrt{3} - 1$

2 $F(x) = 1 - (1 - x)^3, \quad 0 \leqslant x \leqslant 1; \quad 0.206; \quad 0.279$

3 $F(x) = \dfrac{(x - 1)}{x}, \quad 1 \leqslant x \leqslant 3,$

$F(x) = \dfrac{(x + 3)}{9}, \quad 3 < x \leqslant 6; \quad 2; \quad \dfrac{29}{12}$

4 $F(x) = \dfrac{(x - 1)}{2x}, \quad 1 \leqslant x \leqslant 3,$

$F(x) = \dfrac{(x + 3)}{18}, \quad 3 < x \leqslant 15; \quad 6; \quad \dfrac{17}{2}$

5 $F(x) = 3x^2 - 2x^3, \quad 0 \leqslant x \leqslant 1; \quad \dfrac{1}{2}$

Miscellaneous Exercise 8

1 $\dfrac{2}{3}, 2 - \sqrt{2}$

2 $\dfrac{49}{6}, \dfrac{2}{17}$

3 $2, \dfrac{3}{4}; \dfrac{1}{5}$

4 (i) $\dfrac{2}{5}$ (ii) $\dfrac{13}{5}$ (iii) $\dfrac{3}{2}$

5 $\lambda; \quad 0.632$

6 (i) $\dfrac{2}{75}$ (ii) $\dfrac{70}{9}$ (iii) $\dfrac{12}{25}$

7 $\dfrac{3}{32}, 2;$ (i) $\dfrac{5}{32}$ (ii) $\dfrac{27}{32}$

8 $\dfrac{8}{15}, \dfrac{11}{225}; \dfrac{1}{\sqrt{2}}$

9 (i) $\dfrac{1}{25}$ (ii) $5, \dfrac{25}{6}$ (iii) £2·14

10 (i) $\dfrac{1}{50}$ (ii) $\dfrac{20}{3}, \dfrac{50}{9}$ (iii) $\dfrac{(20n + 11)}{10000}, 0.101$

11 $\dfrac{5}{3}, \dfrac{25}{18};$ (i) 0.19 (ii) 0.469 (iii) 0.0711 (iv) 1.9, 1.539

12 (i) 0.68 (ii) 0.62 (iii) 0.30 (iv) 0.64 (v) 0.04; £7·05

13 $3, \dfrac{6}{5};$ 837.5 gallons

15 (i) $\dfrac{2\lambda}{3}, \dfrac{\lambda^2}{18}$ (ii) 16, £3

14 $\dfrac{5}{16}$

16 $\dfrac{7}{10}, \dfrac{2}{3}$

17 (i) $\dfrac{7}{4}, \dfrac{43}{80}$ (ii) $F(x) = \dfrac{x^2(6 - x)}{27}, 0 \leqslant x \leqslant 3$ (iii) 0.927

18 (i) 97.72p (ii) 82.92p

19 1.26

20 0.588, 117

Chapter 9

Normal distribution

Exercise 9.1

1 **a** 0.7881   **b** 0.9147   **c** 0.6826   **d** 0.9715

2 **a** 0.0808   **b** 0.0465   **c** 0.0271   **d** 0.2492

3 **a** 0.3085   **b** 0.1292   **c** 0.0588   **d** 0.0126

4 **a** 0.9918   **b** 0.9671   **c** 0.6387   **d** 0.8466

5 **a** 0.0128   **b** 0.0719   **c** 0.8673   **d** 0.9500

6 **a** 1.63   **b** 1.555   **c** -1.95 **d** -1.037

7 **a** 1.15   **b** 0.74   **c** -0.23 **d** -1.175

8 **a** 1.645   **b** 1.282   **c** -1.96 **d** -2.326

Exercise 9.2

1 **a** 0.9772   **b** 0.2426   **c** 0.1056   **d** 0.6915   **e** 22.84

2 **a** 0.7257   **b** 0.4495   **c** 0.2119   **d** 0.2119   **e** -3.26

3 **a** 0.9192   **b** 0.8400   **c** 0.0113   **d** 0.1003   **e** 97.88

4 **a** 0.9772   **b** 0.3221   **c** 0.9772   **d** 0.0912   **e** 109.36

5 **a** 0.8575   **b** 0.9387   **c** 0.0544   **d** 0.9092   **e** 74.12

Exercise 9.3

1 0.0478

2 0.6730

3 **a** 0.8185   **b** 1.0309

4 70, 40;   **a** grade A **b** fail

5 47.92 min, 7.34 min

6 16.51%, 3.150 cm^2

7 **a** A **b** B

8 £1·06   **9** (i) 0.106   (ii) 219

10 (i) 0.2266   (ii) 54.68   (iii) 78.7 cm^2

11 (i) B (ii) B; A

Exercise 9.4

1 **a** 0.0791   **b** 0.8849   **c** 0.8370

2 **a** 0.1122   **b** 0.1015   **c** 0.6642

3 **a** 0.0441   **b** 0.4264   **c** 0.2868

4 0.8980

5 153 (174 without continuity correction)

Miscellaneous Exercise $\boxed{9}$

1 0.0287

2 (i) 0.317   (ii) 0.337;   0.041

3 0.9155

4 68.71, 5.30; 64.11; 13.3%

5 a 0.0912   (i) 20.07   (ii) 0.02   **b** 0.2420, 0.0269

6 $\sum_{r=50}^{100} \binom{100}{r}(0.42)^{100-r}(0.58)^r$; 0.9575

7 (i) 0.893   (ii) 5380   (iii) 0.0320

8 (i) 0.1587   (ii) 0.0794   (iii) 4.78

9 (i) 0.00122   (ii) 0.0081   (iii) 0.788

10 60.0, 0.2; 9%

11 (i) 8.13 a.m.   (ii) 7

12 a £8·44   **b** £16·86

13 6.43 p.m.

14 (i) 0.055   (ii) 0.0197

15 (i) 0.2048   (ii) 0.05792;   0.2480

16 (i) 0.827   (ii) 14.57   (iii) 707.1 cm^2

17 (i) 0.605   (ii) 0.347

18 0.9155; 0.8, $\dfrac{0.16}{n}$; 153 (174 without continuity correction)

19 0.0954, 0.0004; 286 (271 without continuity correction)

20 0.04

Chapter 10

Revision

Exercise $\boxed{10.1}$

1 a 9, 8   **b** 11, 9.38   **c** median; mean distorted by atypical value 41

2 b 1.5, 1.55   **c** 1

3 7, 7; £90, £26·46

4 b 5, 2.993   **c** 4.79   **d** 8.52

5 (ii) 46600 km   (iii) 46538 km, 20625 km

Exercise $\boxed{10.2}$

2 a $\dfrac{7}{30}$ **b** $\dfrac{5}{14}$

3 $\dfrac{9}{14}$; 22

4 (i) $\dfrac{4}{15}$ (ii) $\dfrac{4}{15}$ (iii) $\dfrac{28}{435}$ (iv) $\dfrac{2}{145}$ (v) $\dfrac{99}{1015}$ (vi) $\dfrac{48}{203}$ (vii) $\dfrac{676}{1015}$

5 (i) 0.12   (ii) 0.184   (iii) 0.32   (iv) 0.25

Exercise 10.3

1 a $2, \dfrac{8}{11}$; $2, \dfrac{32}{11}$ **b** $2, \dfrac{4}{3}$

2 a 0.729, 10   **b** 25

3 $n, \dfrac{(n^2 - 1)}{3}$

4 0.24, 0.0604

5 a 0.416   **b** 0.121   **c** 0.105   **d** 38   **e** 16

Exercise 10.4

1 c 0 (symmetry)   **d** $\dfrac{1}{2}$

2 a 1   **b** $\dfrac{\pi}{2} - 1, \pi - 3$ **c** $\dfrac{1}{2}$

3 a $F(x) = \dfrac{(x - 4)^2}{12}, \quad 4 \leqslant x \leqslant 6$ **b** 6.5   **c** $4 + \sqrt{3}$

    $F(x) = \dfrac{(x - 5)}{3}, \quad 6 < x \leqslant 8$

4 a 0.008   **b** 0.031   **c** 2496

5 a $\dfrac{2}{3}$ **b** $\dfrac{4r}{\pi}$

Exercise 10.5

1 a 0.1587   **b** 0.00063   **c** 0.0228

2 first machine

3 (i) 0.960, 0.788   (ii) 0.857   (iii) 99%

4 0.0432, random sample from a large population, responses truthful

5 a 0.3968   **b** 0.7498   **c** 77,123   **d** biased die or unlikely event

Miscellaneous Exercise 10

1 (i) 46   (ii) 37

2 0.055; 507, $\dfrac{145}{7}$; 0.062

3 3.6, 1.54;   (i) 0.572   (ii) 6

4 1.32, 1.0701, 1.0345;   1.3

5 (i) 1011, 159, 1009   (ii) 0.633

6 $\dfrac{5}{54}, \dfrac{65}{72}$

7 (i) $\dfrac{8}{9}$ (ii) $\dfrac{1}{9}$ (iii) $\dfrac{1}{9}$ (iv) $\dfrac{1}{63}$ (v) $\dfrac{13}{63}$ (vi) $\dfrac{50}{63}$ (vii) $\dfrac{2}{63}$

8 (i) $\dfrac{2}{5}$ (ii) $\dfrac{1}{2}$ (iii) $\dfrac{4}{5}$

9 (ii) $\dfrac{8}{15}$ (iii) $\dfrac{3}{4}$ (iv) $\dfrac{1}{15}$ (v) $\dfrac{5}{13}$

10 (i) $\dfrac{27}{40}$ (ii) $\dfrac{1}{81}$ (iii) 5

11 (i) 0.029   (ii) 0.66

12 (i) $\dfrac{11}{120}$ (ii) $\dfrac{1}{4}$ (iii) $\dfrac{23}{60}$

(iv)

x	3	4	5	6	7	8
p	$\dfrac{10}{120}$	$\dfrac{30}{120}$	$\dfrac{35}{120}$	$\dfrac{31}{120}$	$\dfrac{11}{120}$	$\dfrac{3}{120}$

(v) 12

13 a 0.196   **b** 14

14 $\dfrac{(n+1)}{2}$

15 (i) $P(X = 0) = \dfrac{(N-4)(N-5)(N-6)}{N(N-1)(N-2)}$ $P(X = 1) = \dfrac{12(N-4)(N-5)}{N(N-1)(N-2)}$

$P(X = 2) = \dfrac{36(N-4)}{N(N-1)(N-2)}$ $P(X = 3) = \dfrac{24}{N(N-1)(N-2)}$

(iii) $8 < N < 18$

16 (ii) 85p

17

m	0	1	2	3	4	5	6
p	$\dfrac{64}{15625}$	$\dfrac{576}{15625}$	$\dfrac{2160}{15625}$	$\dfrac{4230}{15625}$	$\dfrac{4860}{15625}$	$\dfrac{2916}{15625}$	$\dfrac{729}{15625}$

$E[M] = \dfrac{18}{5}$,   (i) $\dfrac{7120}{15625}$ (ii) $\dfrac{3645}{15625}$;   2

18 (i) 0.887   (ii) 0.989

19 (i) 0.032   (ii) 0.0406   (iii) 0.2669   (iv) 4.4, 1.48

20 $\dfrac{8}{3}$ or 4

21 (ii) $\dfrac{\pi}{2}$ (iv) 1.855

22 0.05, 0.323

23 $f(x) \geqslant 0$ for all x and $\displaystyle\int_{\alpha}^{\beta} f(x)\,dx = 1$;   12; $\dfrac{2}{5}, \dfrac{1}{25}$; $\dfrac{1}{3}$

24 (i) 2.875 kg   (ii) £4·75, $\dfrac{3}{16}$

25 (i) £5400   (ii) £270, 100

26 a $\dfrac{3}{250}, \dfrac{55}{8}$ **b** (i) 320   (ii) $\dfrac{99}{125}$

27 0.37

28 (i) 0.4   (ii) 0.5   (iii) 0.375   (iv) 0.701   (v) 0.70, 0.01

29 a 1529   **b** 0.3300   (i) B(100, 0.33), 0.33, 0.002211   (ii) 0.0835

30 0.069; 0.548

31 0.96

32 1.67

Statistical tables

Table 1

Cumulative binomial probabilities

The tabulated value is $P(X \leqslant r)$, where X has a binomial distribution with parameters n and p.

	$p =$	0.05	0.10	0.15	0.20	0.25	0.30	0.35	0.40	0.45	0.50
$n = 5$	$r = 0$	0.774	0.590	0.444	0.328	0.237	0.168	0.116	0.078	0.050	0.031
	1	0.977	0.919	0.835	0.737	0.633	0.528	0.428	0.337	0.256	0.187
	2	0.999	0.991	0.973	0.942	0.896	0.837	0.765	0.683	0.593	0.500
	3	1.000	1.000	0.998	0.993	0.984	0.969	0.946	0.913	0.869	0.813
	4			1.000	1.000	0.999	0.998	0.995	0.990	0.982	0.969
$n = 10$	$r = 0$	0.599	0.349	0.197	0.107	0.056	0.028	0.013	0.006	0.003	0.001
	1	0.914	0.736	0.544	0.376	0.244	0.149	0.086	0.046	0.023	0.011
	2	0.988	0.930	0.820	0.678	0.526	0.383	0.262	0.167	0.100	0.055
	3	0.999	0.987	0.950	0.879	0.776	0.650	0.514	0.382	0.266	0.172
	4	1.000	0.998	0.990	0.967	0.922	0.850	0.751	0.633	0.504	0.377
	5		1.000	0.999	0.994	0.980	0.953	0.905	0.834	0.738	0.623
	6			1.000	0.999	0.996	0.989	0.974	0.945	0.898	0.828
	7				1.000	1.000	0.998	0.995	0.988	0.973	0.945
	8						1.000	0.999	0.998	0.995	0.989
	9							1.000	1.000	1.000	0.999
$n = 20$	$r = 0$	0.358	0.122	0.039	0.012	0.003	0.001	0.000	0.000		
	1	0.736	0.392	0.176	0.069	0.024	0.008	0.002	0.001	0.000	
	2	0.925	0.677	0.405	0.206	0.091	0.035	0.012	0.004	0.001	0.000
	3	0.984	0.867	0.648	0.411	0.225	0.107	0.044	0.016	0.005	0.001
	4	0.997	0.957	0.830	0.630	0.415	0.238	0.118	0.051	0.019	0.006
	5	1.000	0.989	0.933	0.804	0.617	0.416	0.245	0.126	0.055	0.021
	6		0.998	0.978	0.913	0.786	0.608	0.417	0.250	0.130	0.058
	7		1.000	0.994	0.968	0.898	0.772	0.601	0.416	0.252	0.132
	8			0.999	0.990	0.959	0.887	0.762	0.596	0.414	0.252
	9			1.000	0.997	0.986	0.952	0.878	0.755	0.591	0.412
	10				0.999	0.996	0.983	0.947	0.872	0.751	0.588
	11				1.000	0.999	0.995	0.980	0.943	0.869	0.748
	12					1.000	0.999	0.994	0.979	0.942	0.868
	13						1.000	0.998	0.994	0.979	0.942
	14							1.000	0.998	0.994	0.979
	15								1.000	0.998	0.994
	16									1.000	0.999
	17										1.000

Table 2

Cumulative Poisson probabilities

The tabulated value is $P(X \leq r)$, where X has a Poisson distribution with mean a.

$a =$	0.5	1.0	1.5	2.0	2.5	3.0	3.5	4.0	4.5	5.0
$r = 0$	0.607	0.368	0.223	0.135	0.082	0.050	0.030	0.018	0.011	0.007
1	0.910	0.736	0.558	0.406	0.287	0.199	0.136	0.092	0.061	0.040
2	0.986	0.920	0.809	0.677	0.544	0.423	0.321	0.238	0.174	0.125
3	0.998	0.981	0.934	0.857	0.758	0.647	0.537	0.433	0.342	0.265
4	1.000	0.996	0.981	0.947	0.891	0.815	0.725	0.629	0.532	0.440
5		0.999	0.996	0.983	0.958	0.916	0.858	0.785	0.703	0.616
6		1.000	0.999	0.995	0.986	0.966	0.935	0.889	0.831	0.762
7			1.000	0.999	0.996	0.988	0.973	0.949	0.913	0.867
8				1.000	0.999	0.996	0.990	0.979	0.960	0.932
9					1.000	0.999	0.997	0.992	0.983	0.968
10						1.000	0.999	0.997	0.993	0.986
11							1.000	0.999	0.998	0.995
12								1.000	0.999	0.998
13									1.000	0.999
14										1.000

$a =$	5.5	6.0	6.5	7.0	7.5	8.0	8.5	9.0	9.5	10.00
$r = 0$	0.004	0.002	0.002	0.001	0.001	0.000	0.000	0.000	0.000	0.000
1	0.027	0.017	0.011	0.007	0.005	0.003	0.002	0.001	0.001	0.000
2	0.088	0.062	0.043	0.030	0.020	0.014	0.009	0.006	0.004	0.003
3	0.202	0.151	0.112	0.082	0.059	0.042	0.030	0.021	0.015	0.010
4	0.358	0.285	0.224	0.173	0.132	0.100	0.074	0.055	0.040	0.029
5	0.529	0.446	0.369	0.301	0.241	0.191	0.150	0.116	0.089	0.067
6	0.686	0.606	0.527	0.450	0.378	0.313	0.256	0.207	0.165	0.130
7	0.809	0.744	0.673	0.599	0.525	0.453	0.386	0.324	0.269	0.220
8	0.894	0.847	0.792	0.729	0.662	0.593	0.523	0.456	0.392	0.333
9	0.946	0.916	0.877	0.830	0.776	0.717	0.653	0.587	0.522	0.458
10	0.975	0.957	0.933	0.901	0.862	0.816	0.763	0.706	0.645	0.583
11	0.989	0.980	0.966	0.947	0.921	0.888	0.849	0.803	0.752	0.697
12	0.996	0.991	0.984	0.973	0.957	0.936	0.909	0.876	0.836	0.792
13	0.998	0.996	0.993	0.987	0.978	0.966	0.949	0.926	0.898	0.864
14	0.999	0.999	0.997	0.994	0.990	0.983	0.973	0.959	0.940	0.917
15	1.000	0.999	0.999	0.998	0.995	0.992	0.986	0.978	0.967	0.951
16		1.000	1.000	0.999	0.998	0.996	0.993	0.989	0.982	0.973
17				1.000	0.999	0.998	0.997	0.995	0.991	0.986
18					1.000	0.999	0.999	0.998	0.996	0.993
19						1.000	1.000	0.999	0.998	0.997
20								1.000	0.999	0.998
21									1.000	0.999
22										1.000

Table 3

The normal distribution

a *Distribution Function.* The tabulated value is $\Phi(z) = P(Z \leqslant z)$, where Z is the standardised normal random variable, N(0, 1).

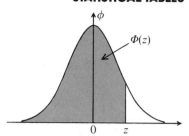

z	.00	.01	.02	.03	.04	.05	.06	.07	.08	.09
.0	.5000	.5040	.5080	.5120	.5160	.5199	.5239	.5279	.5319	.5359
.1	.5398	.5438	.5478	.5517	.5557	.5596	.5636	.5675	.5714	.5753
.2	.5793	.5832	.5871	.5910	.5948	.5987	.6026	.6064	.6103	.6141
.3	.6179	.6217	.6255	.6293	.6331	.6368	.6406	.6443	.6480	.6517
.4	.6554	.6591	.6628	.6664	.6700	.6736	.6772	.6808	.6844	.6879
.5	.6915	.6950	.6985	.7019	.7054	.7088	.7123	.7157	.7190	.7224
.6	.7257	.7291	.7324	.7357	.7389	.7422	.7454	.7486	.7517	.7549
.7	.7580	.7611	.7642	.7673	.7704	.7734	.7764	.7794	.7823	.7852
.8	.7881	.7910	.7939	.7967	.7995	.8023	.8051	.8078	.8106	.8133
.9	.8159	.8186	.8212	.8238	.8264	.8289	.8315	.8340	.8365	.8389
1.0	.8413	.8438	.8461	.8485	.8508	.8531	.8554	.8577	.8599	.8621
1.1	.8643	.8665	.8686	.8708	.8729	.8749	.8770	.8790	.8810	.8830
1.2	.8849	.8869	.8888	.8907	.8925	.8944	.8962	.8980	.8997	.9015
1.3	.9032	.9049	.9066	.9082	.9099	.9115	.9131	.9147	.9162	.9177
1.4	.9192	.9207	.9222	.9236	.9251	.9265	.9279	.9292	.9306	.9319
1.5	.9332	.9345	.9357	.9370	.9382	.9394	.9406	.9418	.9429	.9441
1.6	.9452	.9463	.9474	.9484	.9495	.9505	.9515	.9525	.9535	.9545
1.7	.9554	.9564	.9573	.9582	.9591	.9599	.9608	.9616	.9625	.9633
1.8	.9641	.9649	.9656	.9664	.9671	.9678	.9686	.9693	.9699	.9706
1.9	.9713	.9719	.9726	.9732	.9738	.9744	.9750	.9756	.9761	.9767
2.0	.9772	.9778	.9783	.9788	.9793	.9798	.9803	.9808	.9812	.9817
2.1	.9821	.9826	.9830	.9834	.9838	.9842	.9846	.9850	.9854	.9857
2.2	.9861	.9864	.9868	.9871	.9875	.9878	.9881	.9884	.9887	.9890
2.3	.9893	.9896	.9898	.9901	.9904	.9906	.9909	.9911	.9913	.9916
2.4	.9918	.9920	.9922	.9925	.9927	.9929	.9931	.9932	.9934	.9936
2.5	.9938	.9940	.9941	.9943	.9945	.9946	.9948	.9949	.9951	.9952
2.6	.9953	.9955	.9956	.9957	.9959	.9960	.9961	.9962	.9963	.9964
2.7	.9965	.9966	.9967	.9968	.9969	.9970	.9971	.9972	.9973	.9974
2.8	.9974	.9975	.9976	.9977	.9977	.9978	.9979	.9979	.9980	.9981
2.9	.9981	.9982	.9982	.9983	.9984	.9984	.9985	.9985	.9986	.9986
3.0	.9987	.9987	.9987	.9988	.9988	.9989	.9989	.9989	.9990	.9990
3.1	.9990	.9991	.9991	.9991	.9992	.9992	.9992	.9992	.9993	.9993
3.2	.9993	.9993	.9994	.9994	.9994	.9994	.9994	.9995	.9995	.9995
3.3	.9995	.9995	.9995	.9996	.9996	.9996	.9996	.9996	.9996	.9997
3.4	.9997	.9997	.9997	.9997	.9997	.9997	.9997	.9997	.9997	.9998

b *Upper Percentage Points.* The tabulated value is z_p, where $P(Z > z_p) = p$, so that $1 - \Phi(z_p) = p$.

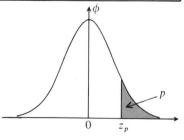

p	0.05	0.025	0.01	0.005	0.001	0.0005
z_p	1.64	1.96	2.33	2.58	3.09	3.29

217

Index

(R) = Revision